76 **Topics in Current Chemistry**
Fortschritte der Chemischen Forschung

Aspects of Molybdenum and Related Chemistry

Springer-Verlag
Berlin Heidelberg New York 1978

This series presents critical reviews of the present position and future trends in modern chemical research. It is addressed to all research and industrial chemists who wish to keep abreast of advances in their subject.

As a rule, contributions are specially commissioned. The editors and publishers will, however, always be pleased to receive suggestions and supplementary information. Papers are accepted for "Topics in Current Chemistry" in English.

ISBN 3-540-08986-1 Springer-Verlag Berlin Heidelberg New York
ISBN 0-387-08986-1 Springer-Verlag New York Heidelberg Berlin

Library of Congress Cataloging in Publication Data. Main entry under title: Aspects of molybdenum and related chemistry. (Topics in current chemistry ; 76) Bibliography: p. Includes index. CONTENTS: Tsigdinos, G. A. Heteropoly compounds of molybdenum and tungsten. – Tsigdinos, G. A. Inorganic sulfur compounds of molybdenum and tungsten – their preparation, structure, and properties. [etc.] 1. Molybdenum compounds–Addresses, essays, lectures. I. Tsigdinos, George A., 1929- Heteropoly compounds of molybdenum and tungsten. 1978. II. Tsigdinos, George A., 1929- Inorganic sulfur compounds of molybdenum and tungsten. 1978. III. Moh, Günter, 1929- High temperature sulfide chemistry. 1978. IV. Series. QD1.F58 vol. 76 [QD181.M7] 540'.8s [546'.534'2] 78-13469

Typesetting and printing: Schwetzinger Verlagsdruckerei GmbH, 6830 Schwetzingen. Bookbinding: Konrad Triltsch, Graphischer Betrieb, 8700 Würzburg
2152/3140–543210

Contents

Springer-Verlag — Postfach 105 280 · D-6900 Heidelberg 1
Telephone (0 62 21) 4 87-1 · Telex 04-61 723

Heidelberger Platz 3 · D-1000 Berlin 33
Telephone (0 30) 82 20 01 · Telex 01-83319

Springer-Verlag New York Inc. — 175, Fifth Avenue · New York, NY 10010
Telephone 4 77-82 00

Heteropoly Compounds of Molybdenum and Tungsten

George A. Tsigdinos

Climax Molybdenum Company of Michigan, Research Laboratory, A Subsidiary of AMAX Inc., Ann Arbor, Michigan 48105, U.S.A.

In Memory of my Wife Shirley

Table of Contents

Introduction

The subject of heteropoly compounds has been covered in many recent reviews. This literature has primarily dealt with the structural chemistry and electronic properties of these complexes and has emphasized not only the heteropoly anions of molybdenum, but also those of tungsten and vanadium[1-3]. Equilibria in solution of heteropoly compounds have also been discussed[4]. In recent years, considerable interest has been shown in the heteropoly compounds of molybdenum not only in the structure and characterization of such compounds but also in those aspects of their chemistry that has made them of importance to industrial applications[5]. General aspects of the structure and properties of heteropoly compounds of molybdenum and tungsten have been presented in an earlier publication[6] and a critical evaluation of preparative procedures of heteropoly compounds has also appeared[7]. The present compilation constitutes an extensive updating of that earlier work[6] with the primary emphasis directed to the preparation, structure, properties, and uses of these compounds. Although only heteropoly compounds of molybdenum will be discussed in detail, some tungsten and other analogs will be included for comparison. No attempt has been made to summarize the complete literature in the field.

The heteropoly electrolytes constitute a large category of coordination-type salts and free acids with each member containing a complex and high-molecular weight anion. In these anions, two to eighteen hexavalent molybdenum (or tungsten) atoms surround one or more central atoms (heteroatoms). Vanadium, niobium, tantalum and transition metals can replace some of the molybdenum (or tungsten) atoms in the heteropoly structure. Typical examples are

$[PMo_{12}O_{40}]^{-3}$, $[SiW_{12}O_{40}]^{-4}$, $[TeMo_6O_{24}]^{-6}$, $[As_2Mo_{18}O_{62}]^{-6}$, $[MnNb_{12}O_{36}]^{-12}$, $[PMo_{10}V_2O_{40}]^{-5}$, and $[SiMo_{11}NiO_{40}H_2]^{-6}$,

where P^{+5}, Si^{+4}, Te^{+6}, As^{+5}, Mn^{+4}, P^{+5} and Si^{+4} are the central atoms or heteroatoms, respectively. Over forty different elements can function as central atoms in distinct heteropoly anions, many of these in more than one series of these anions as will be discussed later.

In 1826 Berzelius first prepared and analyzed the heteropoly compound ammonium 12-molybdophosphate[8]. In 1854 Struve described the heteropoly molybdates of Cr^{+3} and Fe^{+3} but postulated these as double salts[9]. Marignac prepared 12-tungstosilicic acid in 1862 and recognized such compounds as a distinct class rather than double salts[10]. In 1908 Miolati made the first systematic attempt to understand the nature of heteropoly compounds by suggesting a structure for these compounds based on the ionic theory and Werner's coordination theory; Miolati's theory was extensively developed and applied by Rosenheim and his co-workers[11]. Although at present the nature of heteropoly compounds is well understood, these early theories served well as the groundwork for the future.

Classification

Heteropoly compounds may be classified according to the ratio of the number of central atoms to the peripheral molybdenum or other such atoms. Compounds with the same number of atoms in the anion usually are isomorphous and have similar chemical properties. Usually, the heteropolymolybdates and heteropolytungstates containing nontransition elements as central atoms have more structural analogues than those that contain transition elements as central atoms. Table 1 lists all elements

Table 1. Elements capable of acting as central atoms (heteroatoms) in heteropoly compounds

Periodic Group	Element[1]
I	H, Cu^{+2}
II	Be^{+2}, Zn^{+2}
III	B^{+3}, Al^{+3} Ga^{+3}
IV	Si^{+4}, Ge^{+4}, Sn^{+4}(?), Ti^{+4}, Zr^{+4}, Th^{+4}, Hf^{+4}, Ce^{+3}, Ce^{+4}, and other rare earths
V	N^{+5}(?), P^{+3}, P^{+5}, As^{+3}, As^{+5}, V^{+4}(?), V^{+5}, Sb^{+3}(?), Sb^{+5}(?), Bi^{+3}
VI	Cr^{+3}, S^{+4}, Te^{+4}, Te^{+6}
VII	Mn^{+2}, Mn^{+4}, I^{+7}
VIII	Fe^{+3}, Co^{+2}, Co^{+3}, Ni^{+2}, Ni^{+4}, Rh^{+3}, Pt^{+4}(?)

[1]) Some of these elements form heteropoly compounds only with molybdenum or only with tungsten. A question mark after the element denotes doubtful existence of a heteropoly anion.

known to be capable of acting as central atoms in heteropoly compounds. In several cases, the heteropoly compounds reported in the literature have not been characterized. The strychnine salt of the anion $[N^{+5}Mo_{12}O_{40}]^{-3}$ has also been reported[12)] but its existence needs further verification.

Tables 2 and 3 illustrate the principal series of heteropolymolybdates and heteropolytungstates, respectively, which have been reported. Table 4 represents central atoms that form heteropoly anions, the composition and structure of which have not yet been elucidated.

Nomenclature

The nomenclature of heteropoly compounds that has appeared in the literature has been inconsistent. Older designations consisted by prefixing the name of the central atom to the words "molybdate (tungstate)" or "molybdic (tungstic)" acid – for example, "phosphomolybdate" or "silicomolybdate". In addition, Greek prefixes were used to describe the numbers of atoms of the central element and molybdenum or tungsten, *i.e.*, dodecatungstosilicic acid. However, the International Union of Pure and Applied Chemistry (IUPAC) uses a different system[13)]. Names of heteropoly

Table 2. Principal series of heteropolymolybdates

Number of atoms X:W	Principal central atoms	Typical formulas	Central group	Structure by X-ray	Structure shown in Figure	Ref.
1:12	*Series A:* N^{+5}(?), P^{+5}, As^{+5}, Si^{+4}, Ge^{+4}, Sn^{+4}(?), Ti^{+4}, Zr^{+4}	$[X^{+n}Mo_{12}O_{40}]^{-(8-n)}$	XO_4	Known	6, 8, 9	1, 81, 82, 95, 266, 267)
	Series B: Ce^{+4}, Th^{+4}, U^{+4}	$[X^{+n}Mo_{12}O_{42}]^{-(12-n)}$	XO_{12}	Known	14	92)
1:11	P^{+5}, As^{+5}, Ge^{+4}	$[X^{+n}Mo_{11}O_{39}]^{-(12-n)}$	–	Unknown	–	–
1:10	P^{+5}, As^{+5}, Pt^{+4}(?)	$[X^{+n}M_{10}O_x]^{-(2x-60-n)}$	–	Unknown	–	–
1:9	Mn^{+4}, Ni^{+4}	$[X^{+n}Mo_9O_{32}]^{-(10-n)}$	XO_6	Known	15	138)
1:9	P^{+5}	$[X^{+n}Mo_9O_{31}(OH)_3]^{-(11-n)}$	XO_4	Known	–	41)
1:6	*Series A:* Te^{+6}, I^{+7}	$[X^{+n}Mo_6O_{24}]^{-(12-n)}$	XO_6	Known		144, 268, 269)
	Series B: Co^{+3}, Al^{+3}, Cr^{+3}, Fe^{+3}, Rh^{+3}, Ga^{+3}, Ni^{+2}	$[X^{+n}Mo_6O_{24}H_6]^{-(6-n)}$	XO_6	Known	7, 16	266, 78, 58)
2:10	Co^{+3}	$[X_2^{+n}Mo_{10}O_{38}H_4]^{-(12-2n)}$	XO_6	Known	23	181)
2:17	P^{+5}, As^{+5}	$[X_2^{+n}Mo_{17}O_x]^{-(2x-102-2n)}$	–	Unknown	–	–
2:5	P^{+5}	$[X_2^{+n}Mo_5O_{23}]^{-(16-2n)}$	XO_4	Known	–	48, 49)
1m:6m (m unknown)	Co^{+2}, Mn^{+2}, Cu^{+2}, Se^{+4}, P^{+3}, As^{+3}, P^{+5}	$[X^{+n}Mo_6O_x]_m^{-m(2x-36-n)}$	–	Unknown	–	–
4:12	As^{+5}	$[H_4As_4Mo_{12}O_{50}]^{-4}$	Cavity	Known	24	183)
1:1	As^{+3}	$[(CH_3)_2AsMoO_{14}OH]^{-2}$	AsO_4	Known	25	184)

Table 3. Principal series of heteropolytungstates

Number of atoms X:W	Principal central atoms	Typical formulas	Central group	Structure by X-Ray	Structure shown in Fig.	Ref.
1:12	P^{+5}, As^{+5}, Si^{+4}, Ge^{+4}, Ti^{+4}, Co^{+2}, Co^{+3}, Zn^{+2}, Cu^{+1}, Cu^{+2}, Ga^{+3}(?)	$[X^{+n}W_{12}O_{40}]^{-(8-n)}$	XO_4	Known	6, 8, 9	1, 81, 82, 95, 266, 267)
1:10	Si^{+4}, Pt^{+4}	$[X^{+n}W_{10}O_x]^{-(2x-60-n)}$	XO_4	Unknown	–	–
1:9	Be^{+2}	$[X^{+2}W_9O_{31}]^{-6}$	–	Unknown	–	–
1:6	*Series A:* Te^{+6}, I^{+7}	$[X^{+n}W_6O_{24}]^{-(12-n)}$	XO_6	Isomorphous with 6-molybdates	7, 16	144, 268, 269)
	Series B: Ni^{+2}, Ga^{+3}	$[X^{+n}W_6O_{24}H_6]^{-(6-n)}$	XO_6	Isomorphous with 6-molybdates		58, 78, 266)
2:18	P^{+5}, As^{+5}	$[X_2^{+n}W_{18}O_{62}]^{-(16-2n)}$	XO_4	Known	17, 18	74)
2:17	P^{+5}, As^{+5}	$[X_2^{+n}W_{17}O_x]^{-(2x-102-2n)}$	–	Unknown	–	–
2:4:18	X = P^{+5}, As^{+5} Z = Mn^{+2}, Co^{+2}, Ni^{+2}, Cu^{+2}, Zn^{+2}	$[X_2^{+n}Z_4^{+m}W_{18}O_{70}H_4]^{-(28-2n+m)}$	XO_4	Known	Ref. 271)	1)
1m:6m (m unknown)	As^{+3}, P^{+3}	$[X^{+n}W_6O_x]_m^{-m(2x-36-n)}$	–	Unknown	–	–

Table 4. Other species of heteropoly anions[1])

Atomic ratio X:Mo	Atomic ratio X:W	Principal central atoms
1:<6	–	P^{+3}, As^{+3}, Sb^{+3}(?), P^{+5}, As^{+5}, S^{+4}, Se^{+4}, V^{+5}, Co^{+3}
1:>6 (but <12)	–	P^{+5}, Mn^{+4}(?)
–	1:<6	P^{+3}, As^{+3}, Sb^{+3}(?), Al^{+3}, V^{+5}, Mn^{+4}, Bi^{+3}
–	1:>6 (but <12)	Si^{+4}(?), Zr^{+4}, Ti^{+4}, V^{+5}, Sn^{+4}(?), Pt^{+4}(?)

[1]) Many of these species have not been characterized. It is possible that several of these are not true compounds.

anions begin with an Arabic numeral designating the simplest ratio of molybdenum or tungsten atoms to the central atom. This is followed by the prefix "molybdo" or "tungsto" and then by the name of the simple anion (or acid) which contains the central atom in the corresponding oxidation state. In case of ambiguity, Roman numerals may be used to designate the oxidation state of the central atom.

Current knowledge of the structure and properties of heteropoly compounds necessitates a more adequate nomenclature of such compounds by taking into consideration both the structure and degree of polymerization and oxidation of the central atom.

The proposed system of nomenclature is designed to extend the current IUPAC names to describe heteropoly compounds more adequately in cases where information about structure is currently available. In this system, the oxidation state of the central atom is shown by a Roman numeral in parentheses. The prefix molybdo, tungsto, or vanado designates the peripheral atoms, whereas the italicized prefix oct, tet, etc., indicates the stereochemistry (octahedral and tetrahedral) about the peripheral and central atoms. Arabic numerals designate the ratio of the number of peripheral and central atoms. The term dimeric, for example, preceding the name indicates the degree of polymerization of the heteropoly anion, when known. A superscript Arabic numeral at the end of the name indicates the charge of the anion. The Greek letter designates bridging between central atoms.

Examples of nomenclature of the IUPAC and the proposed systems are given in Table 5.

Use of the Literature

Wider use of heteropolymolybdates, in both science and industry, has been hindered by the complexity and confusion of the voluminous literature that has accumulated since Berzelius first observed compounds of this type in 1826. Analyses reported in the older literature are often inaccurate since the atomic and molecular weights are so high that small analytical errors produce great errors in the formulas reported; degradation was often overlooked, and much of the work was unwittingly performed on mixtures. Accordingly, the earlier literature (though often extremely valuable)

Table 5. Nomenclature of heteropoly compounds

Formula	Tentative IUPAC Names[13]	Proposed Names
$Na_3[P^{+5}Mo_{12}O_{40}]$	Trisodium dodecamolybdophosphate(V)	Sodium 12-oct-molybdo-tet-phosphate(V)[3)]
$(NH_4)_6[P_2^{+5}Mo_{18}O_{62}]$	Hexammonium 18-molybdodiphosphate(V)	Dimeric ammonium 9-oct-molybdo-tet-phosphate(V)[6)]
$Na_4[NiW_6O_{24}H_6]$	Tetrasodium hexawolframonickelate(II)	Sodium 6-oct-tungsto-oct-nickelate(II)[4)]
$(NH_4)_6[Co_2^{+3}Mo_{10}O_{38}H_4]$	Hexammonium 10-molybdodicobaltate(III)	Dimeric ammonium 5-oct-molybdo-μ-oct-dicobaltate(III)[6)]
$Cs_3H[SiW_{12}O_{40}]$	Tricesium monohydrogen dodecawolframolsilicate	Cesium molyhydrogen 12-oct-tungsto-tet-silicate(IV)[4)]
$H_4[SiW_{12}O_{40}]$	12-Wolframosilicic acid	12-Oct-tungsto-tet-silicic(IV)[4)] acid
$Na_8[Ce^{+4}Mo_{12}O_{42}]$	Octasodium 12-molybdocerate(IV)	Sodium 12-oct-molybdocerate(IV)[8)]
$K_5[P^{+5}Mo_{10}V_2O_{40}]$	Pentapotassium decamolybdodivanadophosphate	10-Oct-molybdo-2-oct-vanado-tet-phosphate(V)[5)]

should be used carefully and interpreted in the light of more recent findings. Unfortunately, this trend continues to persist (though to a much smaller extent) in modern literature.

Caution is especially necessary when:

1. Formulas are reported for salts of cations that usually precipitate many different species of molybdates, *e.g.*, $CN_3H_6^+$ (guanidinium), Hg_2^{+2}, Ag^+, Cs^+.

2. Analyses were obtained by difference (except in the case of water).

3. Preparations involved conditions that partially decompose heteropolymolybdates. In cases where impure materials were obtained because the precipitation methods used often coprecipitated impurities along with undecomposed heteropoly anion. For example, it was ascertained that addition of silver nitrate to a solution of 12-molybdophosphoric acid yields the desired salt partly decomposed, *i.e.*, the Ag to P ratio was 2.84/1.00 and that of Mo to P 11.18/1.00 rather than 3/1 and 12/1, respectively[5)]. Similarily, addition of four moles of silver nitrate to one mole of sodium 12-molybdosilicate yields only white insolubles whereas the same addition to 12-molybdosilicic acid has produced the pure salt, the effect here being one of pH[5)]. The preparation of free 12-molybdoarsenic acid from water dioxane solutions of sodium molybdate and arsenic acid has been described in the literature[14)] involving the addition of perchloric acid, but it was found[15)] that such procedures lead to high contamination by sodium and by perchlorate ions. The reported preparation of beta-12-molybdosilicic acid from water-ethanol mixtures[16, 17)] could not be reproduced without considerable contamination of the product with sodium perchlorate[15, 18)]. Caution should also be exercised in the preparation of heteropoly compounds by ion exchange techniques as several of these are degraded by the resins.

4. Commercial preparations were accepted as pure without further check.

Systems of Formulation

The literature on heteropolymolybdates uses four systems of writing formulas:

1. Empirical or oxide formulas. These express atomic ratios and oxidation states, but give no structural information. They are still used when structural information is lacking.

2. Miolati-Rosenheim formulas. The elaborate Miolati-Rosenheim theory, now outmoded, dominated the field of heteropoly compounds for several decades, and much of the literature is expressed with these formulas. In Miolati-Rosenheim formulas $[MoO_4]^{-2}$ ions or the now discarded $[Mo_2O_7]^{-2}$ ions are represented as coordinated to the central atoms. The Miolati-Rosenheim formulas are still frequently used deliberately to indicate that modern structural information is lacking.

3. Variants of modern formulas. Some authors indicate whether the central atom is enclosed in a tetrahedron XO_4 or an octahedron XO_6; thus $[GeMo_{12}O_{40}]^{-4}$ is sometimes written $[GeO_4Mo_{12}O_{36}]^{-4}$. Other authors rearrange formulas in different ways to indicate structure; for example, $[Ge(Mo_3O_{10})_4]^{-4}$, shows that four groups of three MoO_6 octahedra each surround the central atom in 12-molybdogermanates.

4. International Union of Pure and Applied Chemistry (IUPAC) formulas. The official system of the IUPAC is little used. In this system, 12-molybdosilicic acid and its sodium salt are written $H_4SiO_4 \cdot 12\ MoO_3 \cdot xH_2O$ and $Na_4SiO_4 \cdot 12\ MoO_3 \cdot xH_2O$. Examples of the first three systems are given in Table 6.

Table 6. Systems of formulation

Modern formula	Empirical formula	Miolati-Rosenheim formula
$Na_3[PMo_{12}O_{40}] \cdot 10\ H_2O$	$3\ Na_2O \cdot P_2O_5 \cdot 24\ MoO_3 \cdot 20\ H_2O$	$Na_3H_4[P(Mo_2O_7)_6] \cdot$ $\cdot\ 8\ H_2O$
$K_4[NiW_6O_{24}H_6] \cdot 9\ H_2O$	$2\ K_2O \cdot NiO \cdot 6\ WO_3 \cdot 12\ H_2O$	$K_4H_6[Ni(WO_4)_6 \cdot 9\ H_2O$
$H_6[As_2Mo_{18}O_{62}] \cdot 35\ H_2O$	$As_2O_5 \cdot 18\ MoO_3 \cdot 38\ H_2O$	$H_{12}[As_2O_2(Mo_2O_7)_9] \cdot$ $\cdot\ 32\ H_2O$

Typical Properties

Many heteropolymolybdates and heteropolytungstates fall into distinct series with properties that differ somewhat from one series to another. However, the heteropoly compounds as a class show the following general properties:

1. Heteropolymolybdates generally have very high molecular weights for inorganic electrolytes, ranging to over 4000.

2. Free acids and most salts of heteropoly anions are extraordinarily soluble in water and are often very soluble in several organic solvents as well.

In water: Most free acids are generally extremely soluble (up to 85% by weight of solution). In general, the heteropoly salts of small cations, including those of many heavy metals, are also very soluble. Usually the larger the cation, the less soluble its salt with a given heteropoly anion. Cs^+, Ag^+, Tl^+, Hg^{++}, Pb^{++}, and the larger alkaline earth salts are often insoluble. The NH_4^+, K^+, and Rb^+ salts of some of the most important heteropoly anions are insoluble, but these three cations form other soluble heteropoly salts. Salts of heteropolymolybdate and heteropolytungstate anions with cationic coordination complexes, alkaloids, or organic amines are usually insoluble. The albumins are coagulated and precipitated by most heteropolymolybdates and heteropolytungstates.

In organic solvents: Many of the free acids and a few of the salts are very soluble in organic solvents, especially if the latter contain oxygen. Ethers, alcohols, and ketones (in that order) are generally the best solvents. The dehydrated salts sometimes dissolve readily in organic solvents; the hydrated salts are insoluble. Both 12-molybdophosphoric acid and its cobalt salt can be dissolved and recovered intact from molten benzoic acid solutions[5)].

3. The crystalline free acids and salts of heteropolymolybdates and heteropolytungstate anions are almost always highly hydrated. A given acid or salt will often form several solid hydrates.

4. Many heteropoly compounds are highly colored, the colors ranging through the spectrum and occurring in many shades.

5. Some heteropoly compounds – and especially heteropolymolybdates – are strong oxidizing agents and can be very readily changed to fairly stable, reduced heteropolymolybdates. The reduction products are colored an intense deep blue. In solution the blue substances obey Beer's Law of Light Absorption. The reduced products can in turn act as reducing agents, and the original colors of the anions are restored on oxidation.

6. Recent work has shown that the free heteropolymolybdic acids are strong acids[15)]. The acids are always stronger than molybdic acid or the simple acid containing the central atom in a corresponding oxidation state.

The free acids generally have several replaceable hydrogen ions. Accordingly, numerous crystalline acid salts have been isolated. The several replaceable hydrogen ions of the acid are typically strong and differ little in dissociation constant. Neutralization of successive hydrogen ions therefore proceeds simultaneously when hydroxyl ion is added to the solution, and breaks between successive hydrogen ions are not detectable in the neutralization curves. (Such curves generally show breaks corresponding to the beginning and end of degradation reactions of the complex anion by hydroxyl ion.) However, these breaks usually occur after neutralization of the replaceable hydrogen ions.

7. All heteropolymolybdate and heteropolytungstate anions are decomposed by strongly basic solutions.

The final products are simple molybdate or tungstate ions and either an oxyanion or a hydrous metal oxide of the central atom:

$$[P_2Mo_{18}O_{62}]^{-6} + 34\ OH^- \longrightarrow 18\ MoO_4^{-2} + 2\ HPO_4^{-2} + 16\ H_2O$$

$$[NiW_6O_{24}H_6]^{-4} + 8\ OH^- \longrightarrow 6\ WO_4^{-2} + Ni(OH)_2 + 6\ H_2O$$

A limited number of heteropolymolybdates exist only in very acidic solutions. However, many exist in nearly neutral solutions, and some in neutral and even slightly basic solutions. Heteropolytungstates are more stable in acid solutions than the corresponding molybdates. As a rule, heteropolytungstates are hydrolytically more stable than the heteropolymolybdates.

If hydroxyl ions are progressively added to a solution containing a given heteropoly anion, the pH generally rises steadily. The anion retains its identity throughout a range of pH until the pH of degradation for that particular anion is reached. Thereafter, the pH generally changes little as more hydroxyl ion is added until the heteropoly anion is either converted to another species that is stable in a higher pH range or else it is completely degraded to simple ions.

8. Throughout specific ranges of pH and other conditions, most solutions of heteropolymolybdates and heteropolytungstates appear to contain predominantly one distinct species of anion. It is generally reasonable to assume that this predominant species is identical with the anion existing in the solid state, in equilibrium with the solution, or is closely related to it; some heteropoly anions are remarkably stable.

Preparation

Heteropolymolybdates are always made in solution, generally after acidifying and heating quantities of reactants.

When the central atom is not a transition element, a soluble molybdate or tungstate may be dissolved with a soluble salt containing the central atom in the appropriate oxidation state. The mixture is then acidified to an appropriate pH range. Sometimes barium molybdate is mixed with a sulfuric acid solution containing the central atom, or molybdenum trioxide is boiled with a solution containing the atom.

When the central atom is a transition metal, a simple salt of that element may be mixed hot with a soluble molybdate or tungstate in a solution of appropriate pH. If the central atom must be raised to an unusual oxidation state, persulfate, peroxide or bromine water are often employed; electrolytic oxidation may also be used. Alternatively, freshly precipitated hydrous metal oxides may be boiled in acidic molybdate or tungstate solutions, or coordination complexes may be decomposed in hot molybdate solutions. Free acids are prepared in several ways:

1. by mixing appropriate quantities of the simple acids;
2. by double decomposition of salts (for example sulfuric acid plus a barium salt);
3. by extraction with ether from acidified aqueous solutions[7, 19, 20];
4. by ion exchange from heteropoly salts[7, 21];
5. from mixed or aprotic solvents[7, 22].

A critical evaluation of preparative procedures of heteropoly compounds through 1970 has appeared[7, 271]. Valuable procedures for preparing phosphorus containing heteropoly compounds are given in Ref.[23]. Refs.[8] and [24–27] give procedures for numerous heteropoly compounds but they should be used with caution since some of these are not critical. An extensive review of the literature and references to the preparation of heteropoly compounds through 1955 may also be found in Ref.[28]. Preparations *via* non-ether routes for 12-molybdophosphoric and 12-molybdosilicic acid and several of their metal salts are given in Ref.[5]. Some pertinent comments on the preparative aspects of common heteropolymolybdates are given below to illustrate some of the details of the procedures used and the precautions that need to be taken in preparing pure compounds.

12-Heteropoly Anions

Failure to recognize the hydrolytic instability of 12-molybdophosphoric acid, $H_3[PMo_{12}O_{40}]$, or the equilibria that exist in H_2O H_3PO_4 molybdate solutions has lead to the development of incorrect preparative procedures for this type of heteropoly compounds. For example, a commercial form of molybdophosphoric acid labeled as $P_2O_5 \cdot 20\ MoO_3 \cdot xH_2O$ is a mixture consisting primarily of the 12-acid[7].

No heteropoly species containing Mo/P ratio of 10/1 exists[29]. This composition was incorrectly assumed to be the 11-acid, $H_7[PMo_{11}O_{39}] \cdot xH_2O$, in physicochemical studies of this type of compound[30]. Reports in the literature[31, 32] that 12-molybdophosphoric acid can be prepared by boiling molybdenum trioxide in phosphoric acid were found to be incorrect and the reaction yielded a crude product containing Mo/P atomic ratios 11/1[5]. The pure product can be obtained only after recrystallization of the crude acid from water at room temperature[5]. 12-molybdophosphoric acid can also be prepared as a yellow crystalline solid by ether extraction of acidified solutions of sodium molybdate and phosphate[23] but the crude product thus obtained must be recrystallized from water at room temperature to yield the desired acid, $H_3[PMo_{12}O_{40}] \cdot 30\ H_2O$, as large yellow crystals[7]. Yields of the pure product prepared by this method are usually low.

All too often, the preparation of free heteropoly acids has been reported in the literature where the common-ion effect has been used to isolate the acid. However, such procedures have overlooked the possibility of considerable contamination of the product by other ions. For example, the preparation of beta-12-molybdosilicic acid from water-ethanol solutions has been described[33, 34]. In this procedure, sodium molybdate and sodium silicate solutions containing Mo/Si ratios equal to 12/1 are acidified with perchloric acid and subsequently excess perchloric acid is used to precipitate the purported free beta 12-molybdosilicic acid. The product obtained by this method, however, was not analyzed for sodium or perchlorate ions[33, 34]. It was found, however, that this method of preparation yields a product contaminated by sodium perchlorate[15, 35]. Similarly, the preparation of free 12-molybdoarsenic acid as described in the literature[14] could not be reproduced[15]. It was found that the addition of perchloric acid to water dioxane solutions containing sodium molybdate and arsenate ions resulted in the formation of yellow solids that contained 1.94–2.83% Na indicating considerable contamination by sodium and perchlorate ions[15]. The free beta-12-molybdosilicic acid has been prepared with negligible sodium contents by passage of acidified (with HCl) sodium molybdate sodium silicate water dioxane solutions through ion exchange to remove sodium ion and subsequent addition of hydrochloric acid to the effluent to precipitate the desired acid[15]. Thus in view of these results and those also obtained when such precipitations are carried out from water ethanol mixtures, it is recommended that the procedures outlined in Ref.[15] for preparing beta-12-molybdosilicic acid by ion exchange be adapted to preparing these and other heteropolyacids free of contaminants from mixed solvents by ion exchange.

The alpha form of the 12-molybdosilicic acid can be prepared in quantitative yield by ether extraction of acidified solutions of sodium molybdate and sodium metasilicate[36)]. A quantitative preparation of this acid has been accomplished by a passage of a solution, prepared by boiling 12 moles MoO_3, 2 moles NaOH and one mole Na_2SiO_3, through a strong cation resin in the acid form[5)].

Although the synthesis of ammonium 12-molybdothorate, $(NH_4)_8[ThMo_{12}O_{42}] \cdot xH_2O$, was reported in the older literature[37)], it could not be duplicated[38)]. The synthesis, however, was carried out by refluxing 25% ammonium paramolybdate solution with 15% $Th(NO_3)_4$[38)]. The product obtained had the composition $(NH_4)_8[ThMo_{12}O_{42}] \cdot 7\ H_2O$. The preparation of the isomorphous 12-molybdouranate anion has been reported. The free acid $H_8[UMo_{12}O_{42}] \cdot 18\ H_2O$ and its ammonium salts have been prepared from aqueous solutions[39)].

9-Heteropoly Anions

The dimeric 18-molybdodiphosphate anion, $[P_2Mo_{18}O_{62}]^{-6}$, is the most typical anion in this series. It is prepared by refluxing for prolonged periods of time acidified solutions containing molybdate and phosphates[23)]. This anion reverts easily to the $[PMo_{12}O_{40}]^{-3}$ in solution, ammonium ion accelerates this change by forming the insoluble $(NH_4)_3[PMo_{12}O_{40}] \cdot xH_2O$[23, 40)]. Recently, the preparation of a new complex of formula $[PMo_9O_{31}(OH)_3]^{-6}$ has been carried out by mild acidification of solutions containing $Na_3[PMo_{12}O_{40}]$, Na_2MoO_4 and NaH_2PO_4. Colorless crystals of composition $Na_3H_3[PMo_9O_{31}(OH)_3] \cdot 10\ H_2O$ have been thus obtained[41)].

6-Heteropoly Anions

The 6-heteropoly compounds of molybdenum have been studied extensively and are well understood. Typical examples are the 6-molybdotellurate anion, $[TeMo_6O_{24}]^{-6}$, and the 6-molybdochromate(III) anion, $[CrMo_6O_{24}H_6]^{-3}$. These are usually prepared by the reaction of ammonium paramolybdate with a soluble salt of the central atom[7)]. Some of these are stable only in very mild acid solution and caution should thus be exercised in preparing them.

12-Heteropolymolybdate Anions with Transition Metals as Peripheral Atoms

Such anions are prepared by converting the 12-molybdate into the 11-molybdate by adjusting the pH of the solution between 4–4.5 with potassium acetate or bicarbonate and then adding the desired divalent or trivalent transition metal salt. Typical examples include the 11-molybdo-1-nickelo(II)silicate anion[42)], $[SiMo_{11}NiO_{40}H_2]^{-6}$. The anion $[SiMo_{11}CrO_{40}H_2]^{-5}$ was prepared by adding the stoichiometric amounts of $Cr(NO_3)_3$ to α-molybdosilicic acid[43)]. Other triheteropoly anions containing transition elements as peripheral atoms like Cu^{+2}, Mn^{+2}, Mn^{+3}, Zn^{+2}, Co^{+2}, Ni^{+2}, and Fe^{+3} have been described[44–46].

Other Heteropoly Anions

The molybdovanadophosphoric acids, $H_4[PMo_{11}VO_{40}]$, $H_5[PMo_{10}V_2O_{40}]$ and $H_3[PMo_9V_3O_{40}]$ can be prepared by ether extraction of acidified solutions of molybdate, vanadate and phosphate[47]. The 10-molybdo-2-niobophosphate anion can be prepared by acidification of solutions containing molybdate, 6-niobate and phosphate[7]. The free acid is unstable in water but stable up to one hour in butanol or butyl acetate. The pyridinium or tetramethylammonium salts of the 10-molybdo-2-niobophosphate anion, $(C_5H_5NH)_4H[PMo_{10}Nb_2O_{40}]$ and $[(CH_3)_4N]_4[PMo_{10}Nb_2O_{40}]$, can be precipitated from the organic phase of the acid by addition of the corresponding ammonium chlorides[7].

Mild acidification of molybdate-phosphate solutions yielded the colorless $[P_2Mo_5O_{23}]^{-6}$ anion, isolated as the sodium salt[48, 49]. The heteropoly acid containing pentavalent antimony, $H_3[SbMo_{12}O_{40}] \cdot 48\ H_2O$, was reported to have been prepared by refluxing M/20 molybdic acid with M/75 potassium pyroantimonate[50], however, more work is necessary to elucidate to exact nature of molybdenum heteropoly compounds containing antimony.

Structure in the Solid

X-ray structural determinations have been made on several heteropoly compounds and also on related isopolymolybdates and tungstates. Structural work on heteropoly compounds in solution, although limited, has been carried out and will be discussed later. For convenience, Tables 2 to 4 show the heteropoly compounds on which structures have been determined. Details of structures will be given under each series or category of heteropoly species.

Representation

The simplest way to represent the anion structures is by polyhedra that share corners, edges or faces with one another. Each Mo or W is at the center of an octahedron, and an oxygen atom is located at each vertex of the octahedron (Figs. 1, 2, and 3). The Mo or W atoms are small and the 0 atoms are large. The ionic radii for Mo^{+6} and W^{+6} are 0.62 and 0.65 A, respectively; that of O^{-2} is 1.40[51].

An MoO_6 or WO_6 octahedron can share corners or edges or both with other MoO_6 or WO_6 octahedron. When two octahedra share an edge, this means that two particular oxygen atoms form part of each octahedron (Figs. 4 and 5). The central atom (also small compared to an oxygen) is similarly located at the center of an XO_4 tetrahedron or XO_6 octahedron. Each such polyhedron containing the central atom

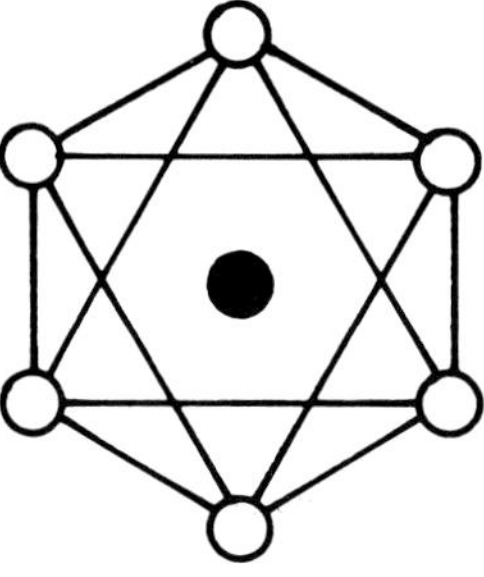

Fig. 1. Locations of the centers of the atoms in an MoO_6 octahedron. The black circle represents molybdenum, the white circles oxygen

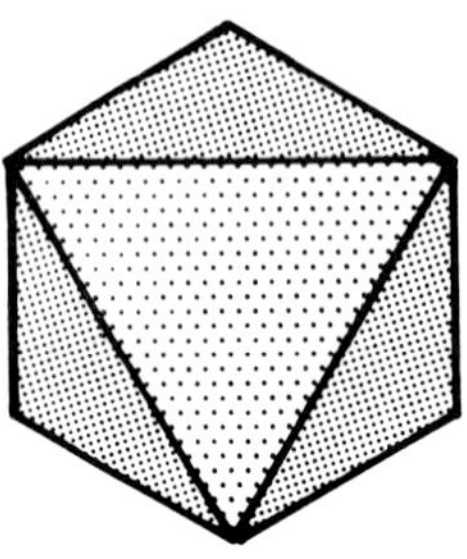

Fig. 2. Diagram of an MoO_6 octahedron. The vertices represent the centers of the six oxygen atoms

Fig. 3. Diagram of an MoO_6 octahedron to the same scale as Figs. 1 and 2 but with the relative sizes of the atoms indicated. The molybdenum atoms is the small black circle

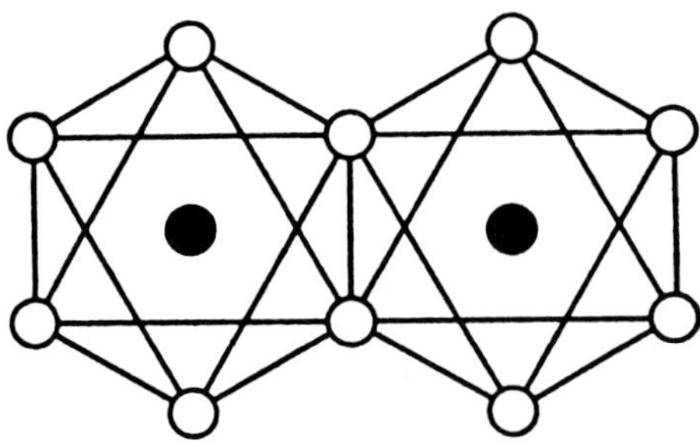

Fig. 4. Two octahedra sharing an edge to form a structural unit Mo_2O_{10}. (This unit does not actually exist.)

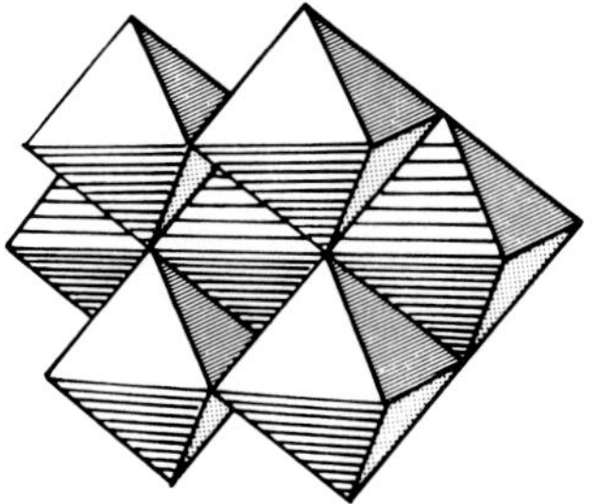

Fig. 5. Structure of the paramolybdate ion $[Mo_7O_{24}]^{-6}$. Here seven MoO_6 octahedra share edges to make the complete unit

is generally surrounded by MoO_6 or WO_6 octahedron, which share corners or edges (or both) with it and with one another so that the correct total number of oxygen atoms is utilized. Each MoO_6 or WO_6 octahedron is directly attached to a central atom through a shared oxygen atom. In the actual structures the octahedra are frequently distorted[1, 52–61].

Another common method of representing structures is by a diagram showing the locations of the centers of the various atoms. Figures 6 (a) and 6 (b) show the 12-molybdate(tungstate) heteropoly anion by this method and by the poly-method. Figure 6 (c) depicts the $[PMo_{12}(W_{12})O_{40}]^{-3}$ anion as a closed polyhedral group of composition $Mo_{12}O_{40}$ or $W_{12}O_{40}$ with the phosphorus atom in the middle.

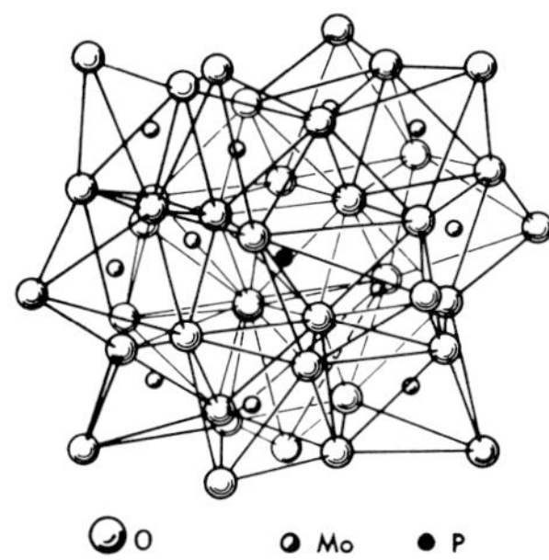

(a) Spatial diagram of the 12-molybdophosphate(V) anion showing the locations of the centers of the various atoms

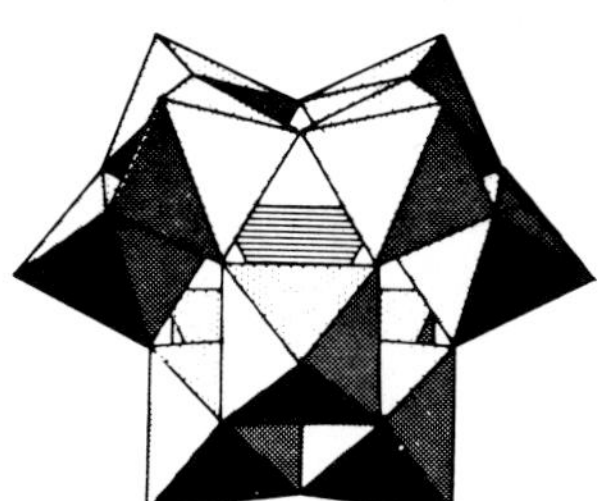

(b) Polyhedral diagram of the 12-molybdophosphate(V) anion to the same scale as Fig. 6 (a); note the central tetrahedron

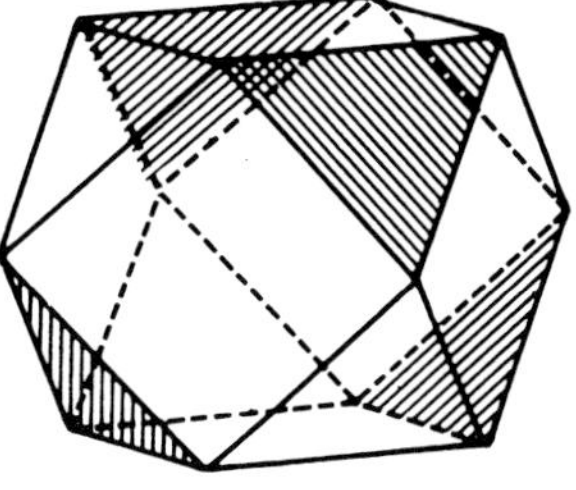

(c) Cubooctahedron diagram of the $[PMo_{12}O_{40}]^{-3}$ or $[PW_{12}O_{40}]^{-3}$ anion; the molybdenum or tungsten atoms are at the corners of the polyhedron and the oxygen atoms are at the midpoints of the edges

Fig. 6. Spatial, polyhedral, and cubooctahedron diagrams of the 12-molybdophosphate(V) anion

These methods of representation do not give pictures of the anions. They are merely diagrams which locate the positions of the centers of the atoms. They do not illustrate the fact that the oxygens are relatively large spheres, as shown in Fig. 3. Therefore, they may give the misleading impression that there are large open spaces within the anions. In reality, practically all of the space within the anion structure is

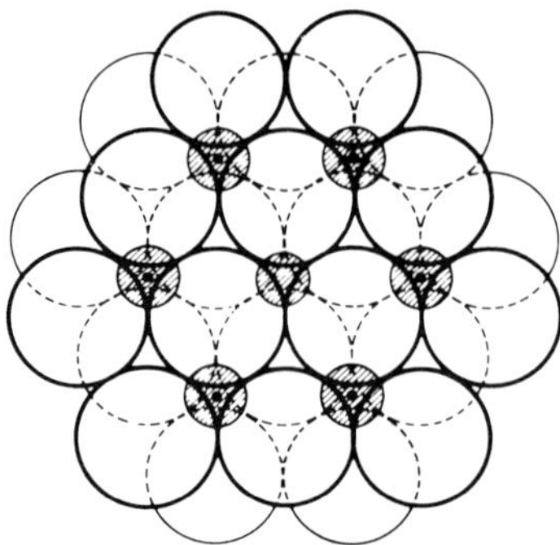

Fig. 7. Diagram of structure of the 6-molybdochromate(III) anion represented with atoms drawn to the scale of their ionic radii (compare with Fig. 16)

taken up by the bulky oxygens, which are either close-packed or nearly so. Figure 7 represents the structure of the 6-molybdochromate(III) anion depicting such packing[40, 58]. The large circles represent oxygen atoms, the six molybdenum atoms are hatched and have black centers. The central atom is hatched and drawn to the scale of the ionic radius of Cr^{+3}. The metal atoms lie in the octahedral interstices between the two layers of close-packed oxygen atoms. The locations of tetrahedral interstices are shown; minor distortions are not indicated in this diagram.

Structure in Solution

X-ray structural analysis of saturated aqueous solutions have shown that the structure of the 12-tungstosilicate anion, $[SiW_{12}O_{40}]^{-4}$, is maintained when the latter is dissolved in water[62, 63]. The same results were obtained for the 12-molybdophosphate anion[63]. Interatomic distances are presented in the references cited. This fact finds further corroboration in the similarity of the UV spectra of solutions of isostructural $H_3[PW_{12}O_{40}]$, $H_4[SiW_{12}O_{40}]$, and $H_6[H_2W_{12}O_{40}]$[64].

The 12-molybdosilicic acid exists in two forms in solution[15, 65–67]. The alpha form (the commonly known compound) is produced in water solution when there are less than 1.5 equivalents of acid per mole of molybdate in the solution during formation. With more than two equivalents of acid per mole of molybdate, the beta form is produced. The two forms have the same empirical formula. The absorption spectra in water are similar, but the extinction coefficient of the beta form is twice as great as that of the alpha modification. Spectroscopic data (Table 7) of the two isomers of this acid, obtained in 1 : 1 dioxane-water solutions, indicate that in this solvent the molar extinction coefficients for the beta-isomer are also roughly twice as large as those for the alpha isomer. However, the reported absorption curves in aqueous solution[65, 67] for the two isomers intersect at wavelengths slightly lower than the values obtained in water dioxane[15]; no such intersection was found in the spectra of the two isomers in water dioxane solutions[15].

It has been suggested that isomers of this kind result from different kind of arrangement of the 12-MoO_6 octahedra about the XO_4 group[68]. However, this geo-

Table 7. Spectroscopic data for $H_4[SiMo_{12}O_{40}]$

Wavelength (NM)	Molar Extinction Coefficient E ($mole^{-1}$ $1 \cdot cm^{-1}$)			
	Alpha-form		Beta-form	
450	1.3		3.9	
425	5.9		12.0	
400		1.9		3.6
375		3.5		7.2
10^5 $H_4[SiMo_{12}O_{40}]$ M	6.4	0.64	6.7	0.66

metrical isomerism in the Keggin structure has been resolved by an X-ray investigation of crystals of beta-$K_4[SiW_{12}O_{40}] \cdot 9\ H_2O$[69, 272]. The $[SiW_{12}O_{40}]^{-4}$ is a geometrical isomer (Fig. 8) of the Keggin structure depicted for the alpha form in Fig. 6. The two structures differ in that in the beta form the upper right Mo_3O_{10} unit is notated by 180°.

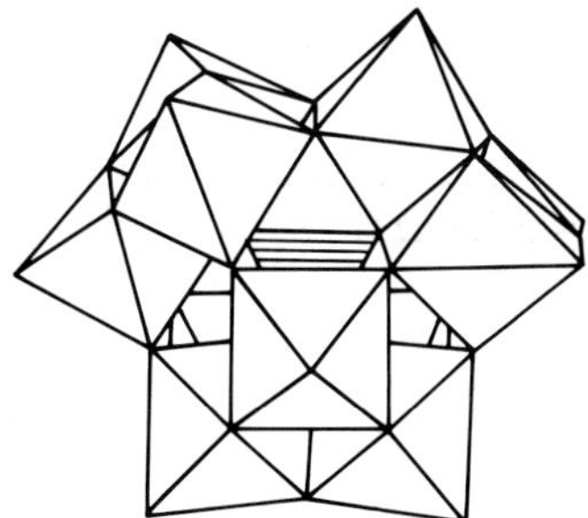

Fig. 8. Model of the β-$[SiW_{12}O_{40}]^{-4}$ anion. Note the rotation of the upper right-hand Mo_3O_{10} unit from that shown in Fig. 6 (b) for the alpha form of the 12-anion

It has been shown that $[SiW_{12}O_{40}]^{-4}$ (ionic weight = 2875) and $[SiMo_{12}O_{40}]^{-4}$ (ionic weight = 1820) have identical diffusion coefficients in solution. These ions are spherical, identical in size and structure[70–72], and have a negligible degree of solvation[72, 73]. Identical diffusion coefficients prove that the thermodynamic radius of these anions (5.6 Å for $[SiW_{12}O_{40}]^{-4}$)[72], is the same as the unsolvated ionic radii in crystals[71], which indicates very low solvation.

Role of Water of Hydration

When the large heteropolymolybdate and heteropolytungstate anions pack together as units in the crystal, the interstices between the anions are very large compared either to water molecules or to most simple cations. In most compounds there is apparently no direct linkage between the individual heteropoly anions as shown in the structure of $K_5[CoW_{12}O_{40}] \cdot 20\ H_2O$[57], and $K_6[P_2W_{18}O_{62}] \cdot 14\ H_2O$[74]. Instead, the complexes are joined by hydrogen bonding through some molecules of water of hydration.

Other water molecules are apparently often zeolitic – that is they are lost on heating continuously rather than stepwise at specific vapor pressures[5, 74]. No great change of crystal structure accompanies the loss of this water. Structures thus obtained may be very porous. The ion exchange properties of various heteropoly salts are attributed to their porous structure and it is believed that ions are able to move freely through the structure so that ion exchange takes place throughout the entire crystal lattice and not only on the surface of the crystals[75–77].

In many cases there are nonzeolitic water molecules, which cannot be lost continuously or without changing the arrangement of the complex ions. This is typified by the 6-molybdochromate(III), $[CrMo_6O_{24}H_6]^{-3}$ (Fig. 7) and 6-tungstonickelate(II), $[NiW_6O_{24}H_6]^{-4}$, anions (shown later in Fig. 16), which contain three molecules of constitutional water[78]. Dehydration experiments by thermogravimetric and differential thermal analysis on $K_3[CrMo_6O_{24}H_6]$ and $K_4[NiMo_6O_{24}H_6]$ have shown that these molecules also contain three molecules of water of constitution[79, 80], in agreement with X-ray studies[58, 78]. In fact, in the structure of $Na_3[CrMo_6O_{24}H_6] \cdot 8\ H_2O$, charge balance and hydrogen-bonding arguments suggest that the six hydrogen atoms of the anion are bonded to the oxygen atoms which are coordinated to the chromium atom[58].

These principles are illustrated in the crystal structure of 12-tungstophosphoric acid hydrate, as determined by X-ray diffraction (Fig. 9)[81, 82]. Of the 29 water molecules, 17 are held together in a well-defined group by hydrogen bonding. The

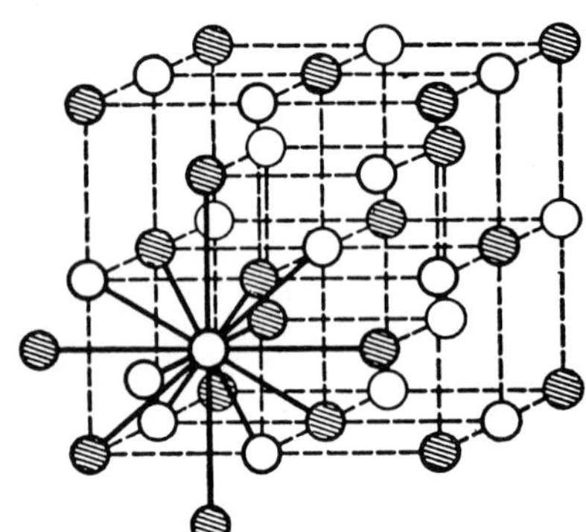

Fig. 9. Structure of crystalline $H_3[PW_{12}O_{40}] \cdot 29\ H_2O$. Open circles represented centers of the $[PW_{12}O_{40}]^{-3}$ units, and shaded circles the centers of the $[H_3 \cdot 29\ H_2O]^{+3}$ units

other 12 are not directly bonded to one another or to any of the first 17. These 12, plus the outer 6 from the first group of 17, are responsible for linking the anions together by hydrogen-bonding. Linnett[56] redescribed this structure by concluding that the 3 H^+ are directly associated with the nonbridged peripheral oxygen atoms of the anion since there is a tendency to be negatively charged. However, a recent nuclear magnetic resonance investigation of the crystals of $H_3[PMo_{12}O_{40}] \cdot 29\ H_2O$ and $H_4[SiMo_{12}O_{40}] \cdot 29\ H_2O$ has shown that the protons are present as H_3O^+ ions[83].

In addition, the above-mentioned compounds showed an IR band at 1730 cm^{-1} indicative of the H_3O^+ ion[83]. Hydronium ions are also present in other heteropoly acids such as $H_6[P_2W_{18}O_{62}] \cdot 16\ H_2O$ and $Na_3H_3[P_2W_{18}O_{62}] \cdot 17\ H_2O$[84–86], as well as in the higher hydrates of 12-molybdoceric(IV) and 12-molybdothoric(IV) acids, namely, $H_8[CeMo_{12}O_{42}] \cdot xH_2O$ and $H_8[ThMo_{12}O_{42}] \cdot xH_2O$[86]. More recently, PMR spectra of hydrates of several heteropoly compounds have been obtained. In 12-molybdosilicic and 12-tungstosilicic acids the PMR spectra show the hydrogen ions in the higher hydrates of the acids to be present as H_3O^+; however, evidence also indicates that there is competition between the anion and water for the proton[87]. The interaction of H^+ and water molecules in the hydrates of the acids $H_6[PW_9V_3O_{40}] \cdot xH_2O$ and $H_7[PW_8V_4O_{40}] \cdot xH_2O$ has been also studied by PMR. In the highly hydrated acids all of the hydrogens are in the form of H_3O^+, but during dehydration the protons become nonequivalent which may account for the possibility of the stepwise basicity of the acids[88].

It should be noted that the protons present in the 6-heteropoly anions like $[CrMo_6O_{24}H_6]^{-3}$ are directly attached to the oxygen atoms bridging the Cr and Mo atoms and are not neutralized prior to the decomposition of these anions with base[89]. In fact, passage of such materials through strong cation exchange resins does not result in the replacement of the anionic protons with the cation of the resin[40].

Heteropoly salts of larger cations, such as cesium, frequently crystallize as acid salts irrespective of the ratio of cations to anions in the mother liquor. Furthermore, salts of these cations are frequently less highly hydrated than those with smaller cations. Apparently, the larger cations take up considerable space between the heteropoly anions; there is little room left for water[58]. In fact, there is often not enough room for all the larger cations required to form a normal salt. Instead, (solvated) hydrogen ions fill in to balance the negative charge of the anions, and a crystalline acid salt results.

12-Heteropoly Anions: Series A

$[X^{+n}Mo(or W)_{12}O_{40}]^{-(8-n)}$

The 12-heteropolymolybdates are apparently subdivided into two series. The structure of Series A anions is built around a central XO_4 tetrahedron and the anion contains 40 oxygen atoms[1]. Series B is typified by the 12-molybdocerate(IV) anion, $[CeMo_{12}O_{42}]^{-8}$, in which there are 42 oxygen atoms present within the anion[89–92] and in which the coordination about the cerium atom is twelve-fold[92]. The 12-molybdothorate(IV) anion may have the same structure.

The Series A 12-acids and their salts are the best known of all heteropolymolybdates and heteropolytungstates and the only compounds of this type used thus far on a large commercial scale.

Tables 2 and 3 show the various central atoms and the type of 12-heteropolymolybdates and 12-heteropolytungstates they form, respectively.

The 12-heteropolymolybdates reported earlier to contain manganese and boron as the central atoms probably do not exist[93, 94].

Crystal Structure

X-ray structure studies have been made with 12-heteropolymolybdates of P^{+5}, As^{+5}, Si^{+4}, Ti^{+4}, and Zr^{+4} [1, 60, 61, 71, 90, 95–99]. Similar studies have been made with the corresponding isomorphous tungstates of B^{+3}, Ge^{+4}, P^{+5}, and Si^{+4} [71]. The anions all have the same symmetrical structure with 12 MoO_6 octahedra or WO_6 octahedra surrounding a central XO_4 tetrahedron. Figures 6(a) and 6(b) represent this structure.

Each oxygen atom of the central XO_4 group is shared with three $Mo(W)O_6$ octahedra. Each $Mo(W)O_6$ octahedron also shares four other oxygen atoms with other $Mo(W)O_6$ groups. The sixth oxygen atom in each $Mo(W)O_6$ group is attached to the molybdenum or tungsten atom alone. These 12 unshared oxygens project from the anion in all directions. Recently, the complete structure of the $[Co^{+3}O_4W_{12}O_{36}]^{-5}$ anion has been determined[57] in the salt $K_5[CoO_4W_{12}O_{36}] \cdot 20\,H_2O$. The work confirms the gross structure proposed by Keggin[95] for the 12-tungstophosphate anion. The anions are arranged in spirals that surround relatively large columns of space. These accommodate potassium ions and zeolitic water. A feature of this structure is the extent to which the tungsten atoms are off-center in their respective WO_6 octahedra. All tungsten atoms are displaced toward the outside of the anion and the peripheral oxygen atoms are drawn somewhat inward. Within each octahedron the longest W–O distance, 2.49 A, occurs between the tungsten and an oxygen atom in the interior of the complex; but the W–O distance involving those peripheral oxygen atoms which are not shared by other tungsten atoms is only 1.43 Å[57]. Infrared evidence based on spectra of 12-heteropolytungstates shows the presence of a peak at 1170 cm^{-1} due to a tungsten-to-oxygen bond[100]. This peak has the highest value thus far observed in any W–O bond, and it is very likely that it may be the result of very short W–O bonds involving peripheral oxygen atoms consistent with the short W–O distances (1.43 Å) obtained by X-rays.

Molecular Weight

All 12-heteropolymolybdates have high molecular weights, generally over 1800, their tungsten analogs over 2800. The anions $[SiMo_{12}O_{40}]^{-4}$ and $[SiW_{12}O_{40}]^{-4}$ have ionic weights of 1819.5 and 2870, respectively. Determination of molecular weights by light scattering of such compounds has been reported in the literature[101].

Color

The 12-heteropolymolybdates are generally yellow in solution and in most crystals. The corresponding tungstates are very pale yellow or colorless. In the 12-series the color also depends on the nature of the central atom. Substituting vanadium for part of the molybdenum in the 12-molybdophosphate anion changes the color from yellow to orange. The Cu^{+}, Ni^{+2}, and Cu^{+2} salts of 12-heteropolymolybdates are green; the Co^{+2} salts are red or brown; and the Fe^{+3} salts may be reddish or yellow. Precipitated salts of basic dyes usually retain the color of the original dye.

Solubility

Solubility of heteropoly compounds in water must be attributed to very low lattice energies and solvation of the cations. Consequently, solubility is governed by packing considerations in the crystals. The cations are fitted in between the spheres of the large negative heteropoly anions. Accomodations of large cations such as Rb^+, Cs^+, Ag^+, and organic bases such as strychnine allow for stable packing in the large interstices and, hence, cause sufficient lowering of the lattice energy to produce insolubility in water.

Although values of lattice energies for heteropoly compounds are not known, estimates have been obtained from the Kapustinskii equation[102, 103]. The lattice energy of $K_4[SiMo_{12}O_{40}]$ was calculated to be approximately 800 kcal/mole[104].

Free 12-acids are remarkably soluble in water (up to 85% by weight of solution) and in dilute acids, alcohols, and ether. For example, 100 ml of water at 25 °C will dissolve 700 g of $H_4[SiMo_{12}O_{40}] \cdot 8\ H_2O$; 100 ml of ethyl acetate, 550 g; and 100 ml of ethyl ether, 400 g. Ether solutions of heteropoly acids are oils insoluble in water and in ether[10, 105] and are believed to be etherates[105]. However, the free acids are insoluble in nonoxygenated solvents such as benzene, chloroform, or carbon disulfide.

Most metal salts are also highly soluble in water. However, rubidium, cesium, mercurous, and thallous salts are insoluble. Ammonium and potassium salts are often insoluble; this property is used in the determination of phosphorus by precipitation as $(NH_4)_3[PMo_{12}O_{40}] \cdot 2\ HNO_3 \cdot xH_2O$.

Insoluble salts also form with many alkaloids, basic dyes, and other organic amines and amides, as mentioned previously.

Similar insoluble salts form with many cationic chelates of heavy metal atoms. Thus, 12-molybdosilicic acid precipitates Cu^{+2}, Ag^+, Cd^{+2}, Zn^{+2}, Sn^{+2}, Cr^{+3}, Ni^{+2}, and Co^{+2} in the presence of ethylenediamine, thiourea, hexamethylenetetramine, dithiooxamide, or similar chelating agents[106].

The solubility of 12-molybdophosphoric and 12-molybdosilicic acids and several transition metal salts has been determined[5] in various solvents and is given in Table 8.

Density of Salts

The density of several heteropoly molybdate salts determined pycnometrically[5] is given in Table 9.

Hydrates

Most free acids of 12-heteropoly anions of Series A form isomorphous 30-hydrates. These melt in their own water of hydration at 40 to 100 °C. They begin to lose water in dry air and give up almost all 30 molecules in vacuum over sulfuric acid. The transition metal salts of the 12-molybdophosphoric and 12-molybdosilicic acids lose water up to 300 °C[5]. The hydrated free acids may be converted to the anhydrous $H_3[PMo_{12}O_{40}]$ and $H_4[SiMo_{12}O_{40}]$ by heating at 180 °C, the products thus obtained can be rehydrated at room temperature[5]. The hydrated acids

Table 8. Solubility data on 12-heteropolymolybdates[1])

Compound	Solubility in water, g/100 ml of Solvent	Solubility in methanol, g/100 ml of Solvent	Solubility in acetone, g/100 ml of Solvent	Solubility in dimethylformamide, g/100 ml of Solvent	Solubility in dimethyl sulfoxide, g/100 ml of Solvent	Solubility in toluene at 25 °C, g/100 ml of Solvent
$Co_3[PMo_{12}O_{40}]_2 \cdot 35\ H_2O$	93.2 (27)	>40 (25)	>100 (27)	– –	– –	Insoluble
$Ni_3[PMo_{12}O_{40}]_2 \cdot 34.5\ H_2O$	85 (27)	186 (25)	113 (26.5)	56 (28)	58 (30)	Insoluble
$La[PMo_{12}O_{40}] \cdot 10\ H_2O$	35 (30)	0.6 (30)	Insoluble	33.5 (30)	<0.5 (30)	Insoluble
$Mn_3[PMo_{12}O_{40}]_2 \cdot 17\ H_2O$	100 (25)	14 (25)	9.2 (25)	70 (28)	86 (30)	Insoluble
$H_3[PMo_{12}O_{40}] \cdot 23\ H_2O$	[2])	~500 (20)	~400 (23)	100 (22)	100 (25)	Insoluble
$Mn_2[SiMo_{12}O_{40}] \cdot 22\ H_2O$	64 (23)	40 (18)	13 (13)	31 (23.5)	72 (27)	Insoluble
$Cu_2[SiMo_{12}O_{40}] \cdot 21\ H_2O$	158 (23)	75 (23)	6 (23)	<5 (23)	12 (23)	Insoluble
$Ni_2[SiMo_{12}O_{40}] \cdot 19.5\ H_2O$	63.5 (24)	24.5 (24)	3 (24)	<4 (24)	17 (25)	Insoluble
$Co_2[SiMo_{12}O_{40}] \cdot 21\ H_2O$	67 (24)	3.5 (24)	2 (24)	<4 (24)	44.5 (24)	Insoluble
$La_4[SiMo_{12}O_{40}]_3 \cdot 82\ H_2O$	~320 (20)	<3 (25)	Insoluble	58 (25)	31 (22)	Insoluble
$H_4[SiMo_{12}O_{40}] \cdot 13\ H_2O$	[2])	– –	– –	~88 (29)	~120 (29)	Insoluble

[1]) From Ref.[5]).
[2]) Solutions contain 85% of the acid by weight are obtained in water.

Table 9. Density of 12-heteropolymolybdate salts

Compound	Density, g/ml
$La[PMo_{12}O_{40}] \cdot 10\ H_2O$	3.673
$Mn_3[PMo_{12}O_{40}]_2 \cdot 41\ H_2O$	2.913
$Ni_3[PMo_{12}O_{40}]_2 \cdot 34\ H_2O$	3.144
$Co_3[PMo_{12}O_{40}]_2 \cdot 34\ H_2O$	3.134
$Cu_2[SiMo_{12}O_{40}] \cdot 20.5\ H_2O$	3.078
$Mn_2[SiMo_{12}O_{40}] \cdot 22\ H_2O$	2.979
$Ni_2[SiMo_{12}O_{40}] \cdot 19.5\ H_2O$	3.080
$Co_2[SiMo_{12}O_{40}] \cdot 24\ H_2O$	3.053
$La_4[SiMo_{12}O_{40}]_3 \cdot 71\ H_2O$	3.009

$H_3[PMo_{12}O_{40}] \cdot 5\ H_2O$, $H_4[SiW_{12}O_{40}] \cdot 5\ H_2O$, and $H_5[BW_{12}O_{40}] \cdot 5\ H_2O$ are isostructural[61).

Salts are nearly all highly hydrated, and numerous isomorphous series exist. Some examples are given in Table 10.

Table 10. Hydrates of 12-heteropolymolybdates

Central atom	Isomorphous hydrate series	Cation M^{+2}
P	$M_3[PMo_{12}O_{40}]_2 \cdot 58\ H_2O$	Mg, Ca, Sr, Ba, Cd, Zn, Mn, Co, Ni
P	$M_3[PMo_{12}O_{40}]_2 \cdot 48\ H_2O$	Ca, Sr, Ba, Cd, Mn, Ni, Co
Si	$M_2[SiMo_{12}O_{40}] \cdot 31\ H_2O$	Cu, Mg, Zn, Mn, Ni, Co
Si	$M_2[SiMo_{12}O_{40}] \cdot 24\ H_2O$	Ca, Sr, Ba

Basicity and Acid Strength

The 12-heteropolymolybdates and 12-heteropolytungstates of Series A all have basicity $(8 - n)$ where n is in the oxidation state of the central atom. Higher basicities that are reported in solution[101, 107)] reflect degradation of the heteropoly anion to simpler species, such as the $H_3[PMo_{12}O_{40}]$ species[4)]. Both the 12-molybdophosphoric acid[108, 109)] and the 12-tungstophosphoric acid[101, 107, 110–112)] have been reported to be heptabasic instead of tribasic. These results arose since both heteropoly anions are extensively hydrolyzed, especially in very dilute solutions[5, 6, 7)]

The strength and degree of basicity of heteropoly acids has been recently elucidated by conductivity and pH measurements. It was shown that the heteropoly acids $H_4[PMo_{11}VO_{40}]$ and $H_5[PMo_{10}V_2O_{40}]$ obey the Onsager-Fuoss theory for strong electrolytes in both aqueous solutions[15, 113)] and in mixed solvents[15, 114)]. The results show that these are strong acids behaving as 1–4 and 1–5 electrolytes, respectively.

Heteropoly acids and salts that undergo partial hydrolytic degradation in water to produce hydrogen ion can be stabilized in mixed solvents, such as water dioxane, water acetone, water alcohol, etc.[15, 114]. Thus, when $H_3[PMo_{12}O_{40}]$ is potentiometrically titrated with base in aqueous solution, it behaves as a six to seven basic acid, however, when similar measurements were carried out in 1 : 1 water-acetone or water dioxane, the acid was found to be tribasic[15, 114].

Data for pH in water and water dioxane solutions show that $H_4[SiMo_{12}O_{40}]$ is a strong acid in both solvents, as shown in Fig. 10, and for $H_3[PMo_{12}O_{40}]$ in Fig. 11. The data show variation of activity coefficients for both acids with increasing concentration[15, 114].

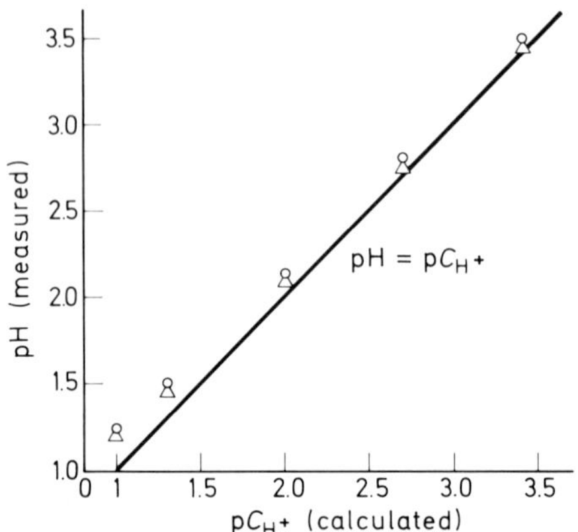

Fig. 10. pH data for $H_4[SiMo_{12}O_{40}]$ in water and 20% dioxane. ○ in water; △ in 20% dioxane

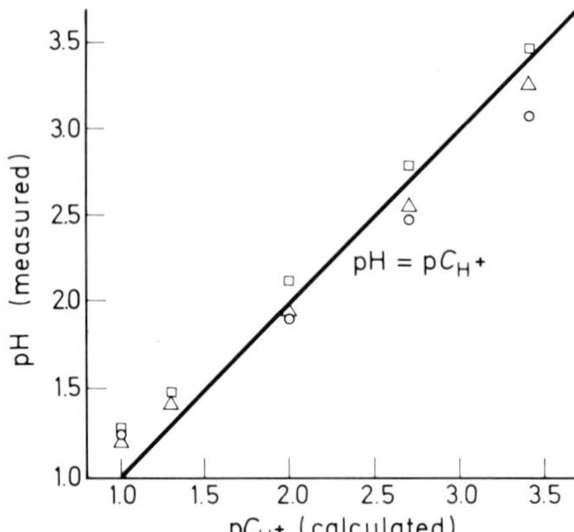

Fig. 11. pH data for $H_3[PMo_{12}O_{40}]$ in water and 20% dioxane. △ Water solution (fresh); ○ water solution after one day; □ 2-% dioxane solution

The pH of 2% solutions of salts of the 12-molybdosilicic and -phosphoric acids is shown in Table 11. The phosphorus compounds have pH values ~2 whereas those of the silicon analog range from 3.2 to 3.7, consistent with the higher hydrolytic instability of the phosphorus compounds[5)].

Table 11. pH of 2% aqueous solutions of salts

Compound	pH of 2% solution
$La[PMo_{12}O_{40}] \cdot 10\ H_2O$	1.84
$Co_3[PMo_{12}O_{40}]_2 \cdot 34\ H_2O$	2.05
$Ni_3[PMo_{12}O_{40}]_2 \cdot 34\ H_2O$	2.01
$Mn_3[PMo_{12}O_{40}]_2 \cdot 41\ H_2O$	2.11
$Co_2[SiMo_{12}O_{40}] \cdot 24\ H_2O$	3.77
$Ni_2[SiMo_{12}O_{40}] \cdot 21\ H_2O$	3.68
$Mn_2[SiMo_{12}O_{40}] \cdot 22\ H_2O$	3.90
$Cu_2[SiMo_{12}O_{40}] \cdot 20.5\ H_2O$	3.72
$La_4[SiMo_{12}O_{40}]_3 \cdot 71\ H_2O$	3.16

Activity coefficients of $H_4[SiW_{12}O_{40}]$ obtained in 0.0004 to 0.04 M concentrations in queous solutions have also been reported[115]. The results obtained agree with the Debye-Hückel theory for 1–4 electrolytes. Osmotic coefficients of 12-tungstosilicic acid have also been reported[116].

Thermal Stability

Thermal stability data on heteropoly compounds often reported in the literature must be viewed with caution since thermal stability limits of such compounds have been assigned on misinterpretation of thermograms. For example, sodium 12-tungstophosphate has been reported to be stable on being heated at 400 °C[117], while 12-tungstosilicic acid was reported to decompose at about 300 °C[118]. 12-molybdophosphoric and 12-molybdosilicic acids have been reported to be stable on heating up to 400 °C[119]. These temperatures, however, are too high and as will be shown later, heteropoly compounds begin to decompose at lower temperatures. Such misinterpretations have arisen by assigning[118–121] the peak temperature in differential thermograms as the decomposition temperature. Such errors occurred in published studies of some 12-heteropolytungstates[118,120,121] as pointed out on studies dealing with the interpretation of dynamic derivative thermogravimetric curves[122]. In fact, the correlation between thermal stability and distance between the central atom and nearest oxygens in isomorphous 12-heteropolytungstates[120] has been shown to be incorrect[122].

Meaningful thermal stability studies on heteropoly compounds must be accompanied by solubility studies of the heated materials to ascertain actual decomposition. In addition, the length of heating must also be specified since such decomposition may be rate dependent in view of the fact that both 12-molybdophosphoric and 12-molybdosilicic acids decompose slowly with time even at room temperature[5]. A detailed study of the thermal behavior of 12-molybdophosphoric and 12-molybdosilicic acids and several of their metal salts has been carried out[5] by the method outlined. The upper thermal stability range obtained for the compounds indicated are shown in Tables 12 and 13, respectively. DTA and TGA data for the two acids in quesion are shown in Figs. 12 and 13. The phosphorus compound is stable up to 350 C whereas the silicon analog begins to decompose slowly above 200 C, the decomposition being complete at 300 C.

Table 12. Thermal stability of 12-molybdophosphates[1])

Compound	Maximum thermal stability (°C)
$H_3[PMo_{12}O_{40}] \cdot 16\ H_2O$	350
$La[PMo_{12}O_{40}] \cdot 10\ H_2O$	300[2])
$Cu_3[PMo_{12}O_{40}]_2 \cdot 25\ H_2O$	350
$Mn_3[PMo_{12}O_{40}]_2 \cdot 17\ H_2O$	300
$Ni_3[PMo_{12}O_{40}]_2 \cdot 34\ H_2O$	300[2])
$Co_3[PMo_{12}O_{40}] \cdot 34\ H_2O$	300[2])

[1]) Samples heated in air at specified temperatures. Heated materials were dissolved in water to ascertain extent of decomposition (Ref. [5])).

[2]) These materials begin to show decomposition at 350 °C.

Table 13. Thermal stability of 12-molybdosilicates[1])

Compound	Maximum thermal stability (°C)
$H_4[SiMo_{12}O_{40}] \cdot 13\ H_2O$	200[2])
$No_4[SiMo_{12}O_{40}] \cdot 12\ H_2O$	250
$Ag_4[SiMo_{12}O_{40}] \cdot 10\ H_2O$	250
$La_4[SiMo_{12}O_{40}]_3 \cdot 71\ H_2O$	250[3])
$Mn_2[SiMo_{12}O_{40}] \cdot 22\ H_2O$	300
$Cu_2[SiMo_{12}O_{40}] \cdot 20\ H_2O$	250
$Ni_2[SiMo_{12}O_{40}] \cdot 21\ H_2O$	300
$Co_2[SiMo_{12}O_{40}] \cdot 24\ H_2O$	250[3])

[1]) Samples heated in air at specified temperatures. Heated materials were dissolved in water to ascertain extent of decomposition (Ref. [5])).

[2]) Acid decomposes slowly above 200 °C by losing constitutional water, the change being complete at 300 °C.

[3]) Partly stable at 300 °C.

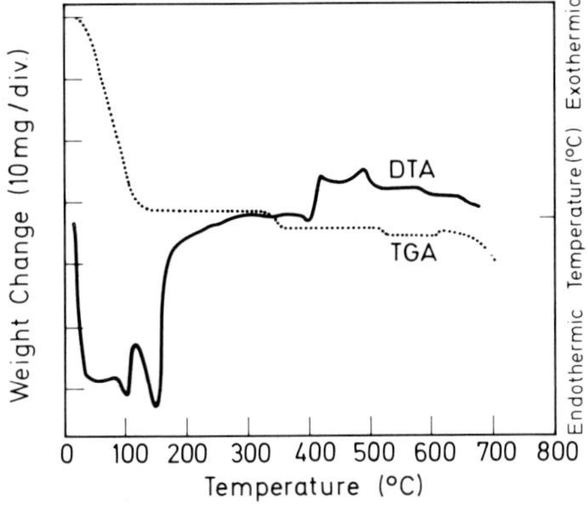

Fig. 12. Thermal behavior of $H_3[PMo_{12}O_{40}] \cdot 16.5\ H_2O$ in nitrogen (sample weight for tga = 0.3062)

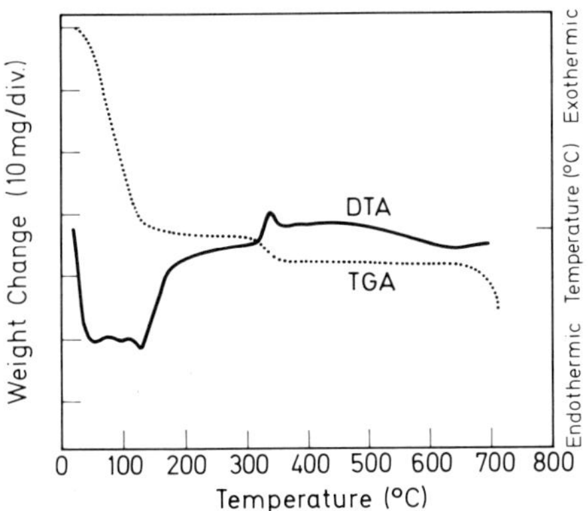

Fig. 13. Thermal behavior of $H_4[SiMo_{12}O_{40}] \cdot 13.5\ H_2O$ in nitrogen (sample weight for tga = 0.2987 g)

Complexes with Organic Compounds

12-heteropoly anions react with many organic oxy-compounds such as sugars, phenols, or acids to give products of unknown structure[23, 36].

Applications in Chemical Analysis

Traces of phosphorus, silicon, and arsenic can be determined colorimetrically by complexing with molybdate and reducing the heteropoly anion to a very deep blue color. Absorption spectra in solution of heteropoly blues show charge transfer bands in the visible region with molar extinction coefficients of the order 10^4 to 10^5 [123], thus they are highly suitable for such work. Phosphate, silicate, and arsenate may be quantitatively separated and determined when they occur together in solution by forming the corresponding heteropolymolybdate complexes and selectively extracting these complexes using organic solvents at the appropriate pH[124].

Oxidation

The ammonium salts of 12-heteropolymolybdate anions (usually insoluble in water) dissolve in solutions of H_2O_2. The oxidized silicon complex is less soluble than the phosphorus one, and the two anions may be separated by using this difference. The structures and formulas of the oxidized products are unknown. A peroxy heteropolymolybdate of composition $(NH_4)H_2[PMo_4O_9(O_2)_2]$ was isolated as a pale yellow solid from a solution of molybdic acid in 30% H_2O_2, addition of H_3PO_4, and precipitation at pH 3–4 by 2-aminopyridine[125].

Degradation

Treatment with alkaline compounds degrades 12-heteropolymolybdates to products with fewer Mo atoms. The first degradation products appear to be 11-heteropolymolybdates and 10-heteropolymolybdates[4]. Further treatment with bases leads to other stable species, mostly having ratios of one X to six or less Mo. In the phosphorus series, the 5-molybdodiphosphate(V) anion and its salts are well established[29, 48, 49]. The crystal structure of these materials is typified by the salt $(NH_4)_5[(MoO_3)_5(PO_4)(HPO_4)] \cdot 3\,H_2O$[126]. Finally, excess alkali causes complete degradation to simple molybdates. The overall degradation requires 20 to 28 moles of NaOH for each mole of 12-acid, the exact number depending upon the valence of the central atom and the chemistry of its simple acid.

The 12-heteropolytungstates of P^{+5} decompose into the 11-series and 10–1/2 series[4], the latter series converts into the 11-series at low pH values. The 12-heteropolytungstates of germanium like the 12-tungstosilicates, are decomposed with alkali to yield 11-tungstogermanates. At high pH, these in turn decompose into tungstates and germanates[127].

Another reaction, similar to degradation, is conversion of the 12-molybdates of P^{+5} to dimeric 9-heteropoly anions by treatment with bases. The conversion appears to proceed through the intermediate 10-heteropolymolybdates.

The different 12-heteropolymolybdates differ markedly in their stability to degradation in water solution. The order of stability is Si > (Zr, Ti) > Ge > P > As. That is, the pH range of stability extends highest for the silicon complex, while very acidic solutions are required to keep the 12-heteropolymolybdates of phosphorus intact as 12-anions. The heptabasicity of 12-molybdophosphoric(V) acid often reported in the literature[11, 107], may be attributed to hydrolytic anionic degradation[5].

The 12-molybdosilicic acid and the corresponding titanium and zirconium 12-acids may be neutralized intact by 4 KOH. Degradation begins only on further treatment with KOH at about pH 4.5 for the Ti and Zr complexes and even then proceeds slowly. No 11-molybdosilicate anion seems to exist; the 12-molybdosilicate anion degrades directly into silicate and molybdate[93]. The 12-heteropolymolybdates of Ti and Zr behave similarly[96].

The 12-molybdogermanic(IV) acid is also quite stable, although degradation sets in before all four replaceable hydrogen ions have been neutralized. The 11-molybdogermanate(IV) anion is stable in the range of pH 3.5 to 4.2[93]. Decomposition into germanate and molybdate starts above pH 4.2 and is complete at pH 5.4.

With 12-molybdophosphoric acid, hydrolytic degradation begins before neutralization of the three acidic hydrogens is complete. The anion is almost certainly intact as $[PMo_{12}O_{40}]^{-3}$ in very acidic solutions[5]. Recent determinations of molecular weight by ultracentrifugation suggest that the anion exists as an 11-molybdophosphate species at pH 4.9[30]. Isolation of the 11-salt, $Na_7[PMo_{11}O_{39}] \cdot xH_2O$, has been claimed.

11-Tungstoborates cannot be obtained by direct degradation of the 12-tungstoborate anion but are prepared directly from acidified solutions of tungstate and borate[4]. They are stable over a limited pH range.

The 12-molybdoarsenates are even less stable than the 12-molybdophosphates. The free acid has not been isolated, and the complex is very easily degraded. The only salts known are the slightly soluble $K_3[AsMo_{12}O_{40}] \cdot 6\,H_2O$ and the difficulty soluble $(NH_4)_3[AsMo_{12}O_{40}] \cdot 6\,H_2O$. In solution, most polymolybdoarsenates are readily converted into dimeric 9-molybdoarsenate anions. With arsenic, these compound types are the most stable and best known heteropolymolybdates and tungstates. It has been shown recently that the 12-molybdoarsenate(V) anion is stable in 1 : 1 water dioxane solutions[15].

In addition to alkaline degradation, 12-molybdophosphates are decomposed by excess of certain acids, such as phosphoric, iodic, periodic, hydrofluoric, and concentrated hydrochloric or sulfuric acid. These acidic decompositions may convert

Table 14. Equilibria in molybdate-phosphate solutions as a function of hydrogen ion concentration (Ref.[4])

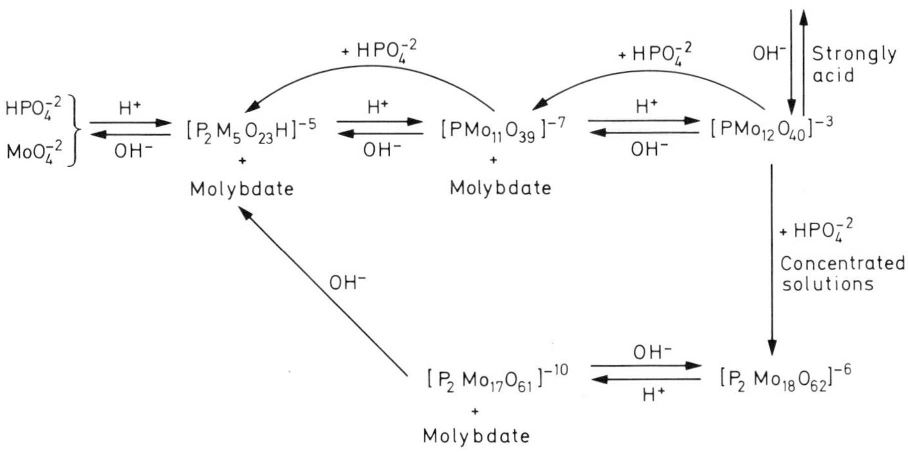

the complexes to heteropolymolybdates with central atoms supplied by the added acid, or to molybdenyl compounds. For example, Te^{+4} displaces silicon from 12-molybdosilicic acid[128].

Molybdenum is of intermediate stability compared with heteropoly anions of other elements. With 12-heteropolytungstates and 12-heteropolyvanadates, the stabilities follow the order W > Mo > V.

The equilibria that exist in solutions of phosphate and molybdate ions in water solution as a function of hydrogen ion concentration are presented in Table 14.

12-Heteropolymolybdates: Series B

$[X^{+n}Mo_{12}O_{42}]^{-(12-n)}$

Central Atom: Ce^{+4}, Th^{+4}, U^{+4}

Preparation

Ammonium 12-molybdocerate(IV), $(NH_4)_8[CeMo_{12}O_{42}] \cdot 8\ H_2O$ is prepared by the addition of ammonium hexanitratocerate(IV) to a boiling solution of ammonium paramolybdate[129, 130]. It can be converted to the acid salt $(NH_4)_6H_2[CeMo_{12}O_{42}] \cdot 12\ H_2O$ by crystallization from 2% H_2SO_4[129]. The free acid $H_8[CeMo_{12}O_{42}]$ can be prepared in solution by ion exchange[130, 131] and isolated as a yellow glass-like mass by prolonged evaporation in air[131]. The corresponding salts of the 12-molybdothorate anion $(NH_4)_8[ThMo_{12}O_{41}] \cdot 8\ H_2O$ and $(NH_4)_6H_6[ThMo_{12}O_{42}] \cdot xH_2O$ are prepared by similar routes[132] but efforts to prepare the free 12-molybdothoric acid by ion exchange were not successful[133]. The free acid $H_8[UMo_{12}O_{42}] \cdot 18\ H_2O$ has been isolated as a yellow solid[39].

Color

The cerium, uranium, and vanadium compounds are yellow. The thorium compounds are white.

Basicity

The basicity of the free acids is $(12 - n)$, where n is the valence of the central atom. Since the known central atoms are in the +4 state, the basicity of all known acids is eight.

Solubility

The free acids and most salts are readily soluble. The normal ammonium and potassium salts are sparingly soluble in cold water, but dissolve in warm acidic solutions. The silver salts are insoluble.

Acid Salts

A series of acid salts exists with the formula $M_6H_2[CeMo_{12}O_{42}] \cdot xH_2O$. A similar series of acid salts exists for the thorium and uranium complexes.

Degradation

The cerium complex can be obtained undegraded in neutral solution[130].

Structural Studies

The Series B of 12-heteropoly compounds is typified by the anion $[CeMo_{12}O_{42}]^{-8}$, the structure of which has been determined in the salt $(NH_4)_6H_2[CeMo_{12}O_{42}] \cdot 12\ H_2O$[92]. This series differs from that of Series A typified by the $[PMo_{12}O_{40}]^{-3}$ anion in that the anion contains 42 oxygen atoms and the central atom is 12-coordinate. An additional feature of this structure involves the first example in which MoO_6 octahedra share common faces. The polyhedron formed by the molybdenum atoms is an icosahedron. Figure 14 depicts the structure of the 12-molybdocerate(IV) anion.

Fig. 14. Idealized sketch of the $[CeMo_{12}O_{42}]^{-8}$ ion showing the linkage of the MoO_6 octahedra

The free heteropoly acids $H_8[CeMo_{12}O_{42}] \cdot 18\ H_2O$, $H_8[ThMo_{12}O_{42}] \cdot 18\ H_2O$ and $H_8[UMo_{12}O_{42}] \cdot 18\ H_2O$ have been studied recently by spectroscopic means[3, 134]. Vibrational spectra confirm the structure found by X-ray[92] for the $[CeMo_{12}O_{42}]^{-8}$ and shown in Fig. 14. This structure is retained in solution as ascertained by Raman spectra[134]. Infrared and proton magnetic resonance spectra indicate that in the acid $H_8[CeMo_{12}O_{42}] \cdot 18\ H_2O$ six protons are present as H_3O^+ whereas the remaining two may possibly form hydrogen bonds between anions. Thus the acid is best formulated as $(H_3O)_6H_2[CeMo_{12}O_{42}] \cdot 12\,H_2O$. The uranium can act as a polydentate ligand without the removal of any of the Mo atoms. Thus it reacts with $ErCl_3$ in solution to form new complexes, the ammonium and cesium salts have been isolated from solution and they have the composition $2.5(NH_4)_2O \cdot 0.66\ ER_2O_3 \cdot UO_2 .\ 12\,MoO_3$ and $1.5\ Cs_2O \cdot 0.67\ Er_2O_3 \cdot UO_2 \cdot MoO_3$[134].
No tungsten analog in this series is known.

11-Heteropoly Anions

$[X^{+n}(Mo\ or\ W)_{11}O_{39}]^{-(12-n)}$

Central Atom: B^{+3}, P^{+5}, As^{+5}, Ge^{+4}

Structure

These compounds are possible degradation products of 12-heteropolymolybdates. Their structure is unknown, but they may be dimeric. The 11-tungstoborate is not obtained by degradation of 12-tungstoborate, nor do the 11-molybdoborates or 12-molybdoborates exist.

Properties

Only the 11-molybdogermanate anion is definitely known[93)]. It exists in solution at pH above 4.8 in equilibrium with germanate, molybdate, and 12-molybdogermanate. Decomposition is complete at pH 5.4. The corresponding 11-tungstogermanates decompose above pH 7.3 into tungstates and germanates[127)].

Recent determinations of molecular weight by ultracentrifugation possibly indicate a monomeric 11-molybdophosphate as the principal species formed by dissolving 12-molybdophosphoric acid in a buffer of pH 4.5[29)]. Other evidence more strongly supports the existence of 11-molybdophosphates in solution[28)]. Solid 11-molybdophosphates have also been reported. The 11-complexes appear to differ in properties from the 12-anions. For example, they are said to be relatively inert to reduction[127)].

Several heteropolytungstates earlier postulated to belong to the 11-Series[135)], were actually found to be 12-heteropolytungstates[121, 123)].

10-Heteropoly Anions

$[X^{+n}Mo_{10}O_x]^{-(2x-60-n)}$

Central Atom: P^{+5}, As^{+5}

Properties

These compounds appear to be degradation products of 12-heteropolymolybdates, but their existence is still controversial. The reported insoluble silver and guanidinium salts of high basicity may not have been correctly formulated because of preparative and analytical difficulties.

The structure of these anions may be dimeric. Solutions and salts of them are reported to be yellow.

Extraction of 10-molybdophosphoric(V) acid with ether gives a system with three layers similar to that obtained with 12-molybdophosphates[137)]. The ether complex layer always contains an atomic ratio P:Mo of 1:10, although the conditions and proportions vary in the other phases. The etherate thus appears to be a distinct compound of approximate composition $H_3PO_4 \cdot 10\,MoO_3 \cdot 20\,Et_2O \cdot 64\,H_2O$. Addition of NaCl to the aqueous phase inhibits formation of the ether complex. The same results are obtained when either HCl or HNO_3 is added to vary the pH.

9-Heteropoly Anions

$[X^{+n}Mo_9O_{32}]^{-(10-n)}$

Central Atom: Mn^{+4}, Ni^{+4}

Structure

The 9-heteropolymolybdates are built around a central XO_6 octahedron, as shown in Fig. 15[138]. Piezoelectric studies and the X-ray determination show that the

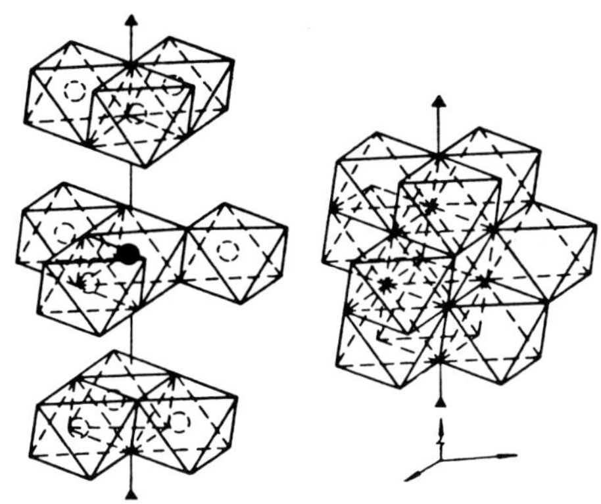

Fig. 15. Structure of the asymmetric anion $[MnMo_9O_{32}]^{-6}$. Left: Exploded view showing manganese atom (black circle) and molybdenum atoms (open circles). Right: View of complete ion

$[MnMo_9O_{32}]^{-6}$ anion is asymmetric. Magnetic and susceptibility studies confirm the oxidation states of the central atoms.

The salts $(NH_4)_6[MnMo_9O_{32}] \cdot 6\,H_2O$ and $(NH_4)_6[NiMo_9O_{32}] \cdot 6\,H_2O$ were shown by X-ray to be isomorphous[139, 140].

Molecular Weight

Molecular weights of 9-heteropolymolybdates are over 1400. Ionic weight of the anion $[MnMo_9O_{32}]^{-6}$ is 1430.5.

The $[MnMo_9O_{32}]^{-6}$ and $[NiMo_9O_{32}]^{-6}$ anions were shown to be monomeric in solution[141].

Color

The 9-molybdomanganate(IV) anion is bright orange-red. The corresponding nickel anion is such a dark red that it is almost black. Its solutions are also very dark.

Basicity

This series has basicities of $(10 - n)$ where n is the valence of the central atom. Since central atoms are all in the +4 state, in practice the basicity is always six.

Solubility

The anions form isomorphous normal potassium and ammonium salts that are slightly soluble in cold water.

Degradation

The 9-heteropolymolybdates and their colors are destroyed in solution by excess of bases or strong acids and by strong reducing agents[142].

The $[MnMo_9O_{32}]^{-6}$ anion is decomposed above pH 5; $[NiMo_9O_{32}]^{-6}$ decomposes above pH 5.5. Unstable $H_6[MnMo_9O_{32}]$ has been prepared in aqueous solution[141].

Preparation

The 9-heteropolymolybdates are prepared by oxidizing a solution of a simple divalent salt of the central atom and a soluble paramolybdate $M_6Mo_7O_{24}$ with persulfate, peroxide, or bromine water[141, 142].

No corresponding tungstates in this series are known.

6-Heteropoly Anions: Series A

$[X^{+n}(\text{Mo or W})_6O_{24}]^{-(12-n)}$

Central Atom: Te^{+6}, I^{+7}

Some compounds of unknown structure listed here with the 1m-6m-heteropolymolybdates may also belong in this group.

Structure

Preliminary structure determination on $(NH_4)_6[TeMo_6O_{24}] \cdot 7\ H_2O$ and the isomorphous $K_6[TeMo_6O_{24}] \cdot 7\ H_2O$ showed that the 6-molybdotellurate(VI) structure consists of seven octahedra all lying in one plane. The six MoO_6 octahedra form a ring surrounding the central TeO_6 octahedron. Each MoO_6 shares one edge with each of its two neighboring MoO_6 octahedra. Each MoO_6 also shares an edge with the TeO_6 octahedron (Fig. 16)[1, 143]. The complete structure of the 6-molybdotellurate anion has now been obtained as it occurs in the compound

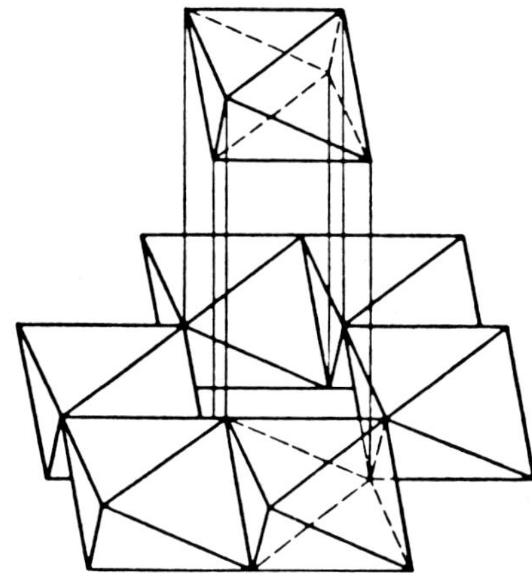

Fig. 16. Structure of the ion $[TeMo_6O_{24}]^{-6}$ with the central TeO_6 octahedron elevated to show the annular arrangement of the six MoO_6 octahedra

$(NH_4)_6[TeMo_6O_{24}] \cdot Te(OH)_6 \cdot 7\,H_2O$, in which it forms a crystal complex with telluric acid[144]. It is isomorphous with the 6-molybdochromate(III) anion[58] (Fig. 16).

The formation of the $[Te^{+6}W_6O_{24}]^{-6}$ anion has also been reported[145]. The corresponding $[I^{+6}W_6O_{24}]^{-5}$ anion is also known[146].

Molecular Weight

The 6-molybdotellurates have molecular weights above 1050. The $[TeMo_6O_{24}]^{-6}$ anion has an ionic weight of 1087.3. The ionic weight of $[TeW_6O_{24}]^{-6}$ is 1685.1.

Color

The 6-heteropolymolybdates are white or yellowish. The corresponding tungstates are white.

Basicity

Basicity of these compounds is $(12 - n)$ where n is the valence of the central atom.

Hydrates

Salts of this series are usually hydrated. Some typical hydrates are:

$Na_6[TeMo_6O_{24}] \cdot 22\ H_2O$

$Li_5[IMo_6O_{24}] \cdot 15\ H_2O$

$Li_5[IMo_6O_{24}] \cdot 9\ H_2O$

$Na_5[IMo_6O_{24}] \cdot 17\ H_2O$

$Na_5[IMo_6O_{24}] \cdot 13\ H_2O$

$Na_5[IW_6O_{24}] \cdot 8\ H_2O$

Solubility

These salts are soluble in water.

Degradation

The 6-heteropolymolybdates are stable at pH values above the stability range for 12-heteropolymolybdates and 9-heteropolymolybdates. However, in neutral or basic solution they are degraded to compounds of lower ratio or to simple salts.

Preparation

The 6-heteropolymolybdates and 6-heteropolytungstates of tellurium are prepared by acidifying solutions containing a tellurate and a molybdate or tungstate, or by adding a base to a mixture of telluric and molybdic acids.

The 6-molybdoiodates or 6-tungstoiodates are prepared similarly. The free acid is made directly from periodate and molybdenum trioxide. It may also be prepared by reaction between barium molybdate, periodic acid, and sulfuric acid. Its neutralization curve has been reported[147]. The yellow free acid evolves gaseous products upon isolation[27] and, therefore, probably undergoes some decomposition.

The formation of complexes between molybdate ions and H_5IO_6 has been studied[148, 149]. The heteropoly complex first forms upon acidification and thereafter stepwise protonation occurs with final formation of $(H_6IMo_6O_{24})^+$.

6-Heteropoly Anions: Series B

$[X^{+n}(\text{Mo or W})_6O_{24}H_6]^{-(6-n)}$

Central Atom: Al^{+3}, Cr^{+3}, Fe^{+3}, Co^{+3}, Rh^{+3}, Ga^{+3}, Ni^{+2} (with molybdenum), Ni^{+2}, Ga^{+3} (with tungsten)

Structure

A combination of magnetic susceptibility studies, X-ray investigations, ionic weights in solution, and chemical evidence has demonstrated that these anions have a monomeric structure based on XO_6 octahedra[1, 79, 89, 92, 150, 151]. Series B is distinguished from Series A in that three molecules of water of constitution are required so that the total number of oxygen atoms within the anion is twenty-four. The 6-tungstonickelate(II) anion has the structure shown in Fig. 16[78]. It has been shown that the six hydrogen atoms in the anion are attached to the six oxygens that bridge the central and molybdenum atoms[58].

Molecular Weight

The 6-heteropolymolybdates have molecular weights above 1000. The $[CrMo_6O_{24}H_6]^{-3}$ anion has an ionic weight of 1017.7. The ionic weight of the $[NiW_6O_{24}H_6]^{-4}$ anion is 1551.8.

Color

Complexes of each central atom have characteristic color in solids and solution, for example, the heteropolymolybdates:

Al – Colorless
Cr – Dark pink
Fe – Colorless
Co – Green-blue
Rh – Amber
Ni – Blue (for tungstate also)
Ga – Colorless (for tungstate also)

Basicity

All 6-heteropoly anions have the basicity of $(6 - n)$ where n is the oxidation state of the central atom.

Hydrates

These complexes are highly hydrated. Water is given up by heating up to the decomposition temperature. For example, DTA and TGA studies have shown[79] that the salt $K_3[CrMo_6O_{24}H_6] \cdot 7\,H_2O$ loses 4 H_2O at 35 °C, 1.5 H_2O at 100 °C, and 1.5 H_2O at 120 °C. The last three H_2O, which are constitutional, are lost at 200 °C, at which point the compound decomposes. The presence of the three constitutional waters have also been confirmed by D_2O experiments[79].

Solubility

The free acids are insoluble in ether, although extremely soluble in water. Nearly all the salts are somewhat soluble in water, but insoluble in organic solvents. The anions precipitate alkaloids, organic amines, cesium, and cationic coordination complexes.

Reduction

The complexes are insensitive to mild reduction, but are decomposed by strong reducing agents such as zinc. The redox properties of these anions will be discussed later.

Degradation

The stability and degradation ranges of pH for the 6-heteropolymolybdates are as follows:

Central Atom	Approximate Stability Range, pH	Approximate Range of Complete Degradation, pH
Al	2–5	5–6
Cr	5.5	5.5–6.5
Fe	2–4.5	4.5–5.5
Co	5	5–6
Ga	5	5–6

Thermal Stability

All of these anions are very stable in solution and as solids, except for the iron and aluminum complexes, which decompose when heated in solution. However, the chromium and cobalt compounds are stable in hot solution, and solutions of free acids may be kept unchanged for long periods of time. These complexes decompose when the anions lose constitutional water at about 200 °C. For example, the salt $K_4[NiMo_6O_{24}H_6] \cdot 5\ H_2O$ loses 5 waters at 80 °C and three constitutional waters at 180 °C[80].

Preparation

The 6-heteropolymolybdates and 6-heteropolytungstates are prepared by mixing hot solutions of paramolybdate or paratungstate ions and simple salts of the central atom[89]. An oxidizing agent such as hydrogen peroxide must be added during the preparation of the cobaltic complex.

The free acids may be prepared from salts by ion exchange and are obtained as solids by evaporating their solutions. Upon evaporation of the pink acid of the chromic complex, a green solid separates, but this redissolves to form a pink solution.

1m-6m-Molybdates

$[X^{+n}Mo_6O_{x\,m}]^{-m(2x-36-n)}$

Central Atom: Cu^{+2}, P^{+3}, P^{+5}, As^{+3}, Se^{+4}, Mn^{+2}, Co^{+2}

Structure

The structure, molecular weights, and degrees of polymerization, m, of these compounds are unknown. Some may be 6-heteropolymolybdates – for example, the P^{+3} and As^{+3} complexes.

Color

These compounds usually have colors characteristic of the central atom. Colors are:

Cu	– Blue	Mn	– Red
P (+3 and +5)	– Yellow or white	Co	– Red
As	– Yellow		

Stability

The Mn complex is quite unstable in solution. Stability of the other complexes is not known.

Preparation

The P^{+5} anions are prepared from molybdic acid and a metaphosphate or pyrophosphate. Manganese, cobalt, and copper form solutions of the divalent cations with paramolybdate anion.

Dimeric 9-Heteropoly Anions

$[X_2^{+n}(Mo\text{ or }W)_{18}O_{62}]^{-(16-2n)}$

Central Atom: P^{+5}, As^{+5}, Be^{+2}

Structure

The 9-heteropolymolybdates and 9-heteropolytungstates are dinuclear complexes containing two central XO_4 tetrahedra surrounded by 18 MoO_6 or 18 WO_6 octahedra. Although the structures of these particular heteropolymolybdates have not been determined by X-ray, the structure of the corresponding dimeric 9-tungstophosphate anion has been characterized in the salt $K_6[P_2W_{18}O_{62}] \cdot 14\ H_2O$, Fig. 17[14]. There is little doubt that the corresponding 9-heteropolymolybdates have the same structure, especially since they have the identical dimeric formula, and the properties and methods of preparation[150] are similar.

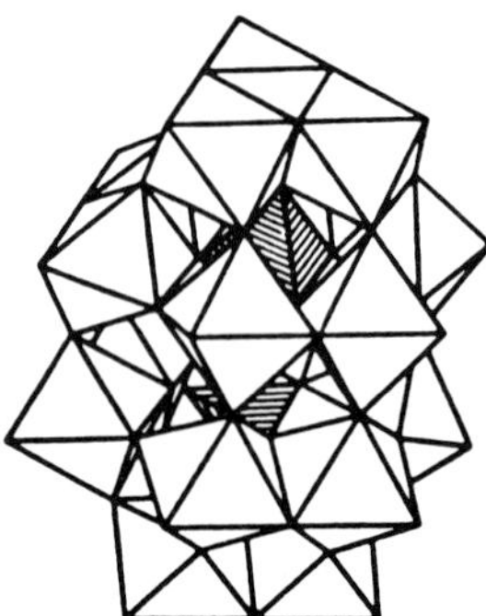

Fig. 17. Structure of the dimeric anions $[P_2W_{12}O_{62}]^{-6}$ and $[P_2Mo_{18}O_{62}]^{-6}$

The half-unit obtained by splitting the dimer has exactly the same structure as the 12-tungstophosphate(V) anion minus three WO_6 groups, Fig. 18. That is, if three adjacent WO_6 groups are removed from a 12-tungstophosphate(V) anion, the half-unit is obtained. Two half-units need only join together to produce the dimeric 9-tungstophosphate(V) anion. This structucal relationship is probably responsible for the easy inter-conversion of the two types of anions and for their many similar properties, especially susceptibility to reduction.

The preparation of a 9-tungstoberyllic(II) acid has also been reported[152]. Although the free acid is difficult to isolate, the guanidinium and lead salts of this acid have been formulated as $3\ (CN_3H_6)_2O \cdot BeO \cdot 9\ WO_3 \cdot 3\ H_2O$ and $3\ PbO \cdot BeO \cdot 9\ WO_3 \cdot 35\ H_2O$. The 9-tungstoberyllate(II) anion may belong to the above-mentioned series with the formula $[Be_2W_{18}O_{62}]^{-12}$.

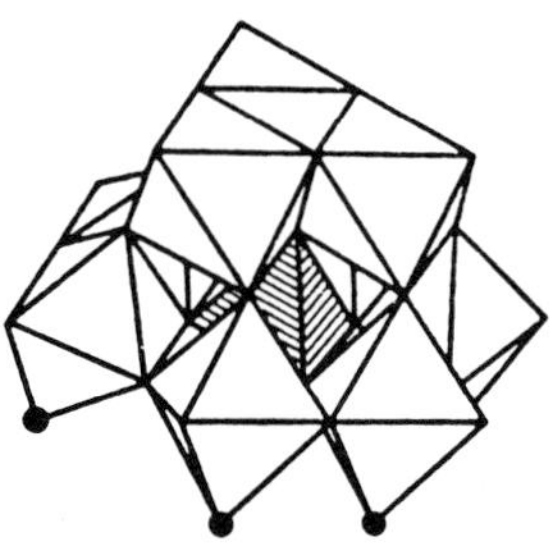

Fig. 18. Half-unit obtained by splitting the $[P_2W_{18}O_{62}]^{-6}$ structure. This half-unit may also be obtained by removing WO_6 octahedra from the 12-heteropolytungstate structure [Fig. 6 (b)]

Molecular Weight

The dimeric 9-heteropolymolybdates have molecular weights above 2750. The ionic weight of the $[P_2Mo_{18}O_{62}]^{-6}$ anion is 2781.0. The $[P_2W_{18}O_{62}]^{-6}$ anion has an ionic weight of 4363.2.

Color

The free acids of molybdenum and their salts are nearly all bright yellow. The analogous tungstates are light yellow. Accordingly these compounds are sometimes given the trivial names "luteophosphomolybdates" and "luteoarsenomolybdates". Some reddish-orange hydrates exist. Nickel and copper salts are green and cobalt salts are brown.

Basicity

The acids are 6-basic. Guanidinium, cesium, silver, and thallium salts of different basicities have sometimes been reported but these are almost certainly mixtures.

Hydration

The crystalline free acids and salts are all highly hydrated. Corresponding compounds of the phosphorus and arsenic series differ in the amounts of water of crystallization. For example, each of the free acids is reported in two forms:

$H_6[P_2Mo_{18}O_{62}] \cdot 33\ H_2O$ – Orange

$H_6[P_2Mo_{18}O_{62}] \cdot 37\ H_2O$ – Bright Orange

$H_6[As_2Mo_{18}O_{62}] \cdot 25\ H_2O$ – Orange-red

$H_6[As_2Mo_{18}O_{62}] \cdot 35\ H_2O$ – Yellow

The higher hydrates of both acids are unstable at ordinary temperatures.

Solubility

The free dimeric 9-molybdophosphoric(V) and 9-tungstophosphoric(V) acids are exceptionally soluble in water (up to 85% by weight of solution), ether, absolute alcohol, and cold nitric acid. However, they are insoluble in nonpolar solvents such as chloroform, carbon disulfide, or hydrocarbons. The 2 : 18 arsenic acid is decomposed by hydrochloric and sulfuric acids.

Ammonium salts of both acids are readily soluble. This property may be used to remove traces of 12-molybdophosphoric acid from 9-molybdophosphoric acid. However, salts of pyridine, alkaloids, and some other organic bases are insoluble.

Most metal salts are readily soluble in water. However, the potassium salt of the 9-molybdophosphoric acid is only slightly soluble in cold water. Rubidium and cesium salts have some solubility. In the arsenic series the lead, mercurous, silver, cesium, and thallous salts are insoluble.

Degradation

The 9-molybdophosphates are apparently degraded by small amounts of hydroxides or carbonates to dimeric 17-molybdodiphosphates of the type $M_{10}[P_2Mo_{17}O_{61}] \cdot xH_2O$. The analogous tungstates yield similar species. Treatment with HCl regenerates the 1:9 acid. Addition of excess base leads to complete decomposition of the anion.

The sensitivity of 12-heteropolymolybdates to reduction may be apparently a direct consequence of the 12-heteropoly structure, since the 11-heteropolymolybdates (formed by degradation of the 12-anions) are insensitive to mild reducing agents. However, the 9-heteropolymolybdates of P^{+5} and As^{+5} are even more sensitive to reduction than the 12-anions. Such reduction usually proceeds stepwise. For example, reduction in the 9-molybdophosphate(V) anion proceeds in definite steps corresponding to the addition of 2, 4, or 6 electrons[23, 158].

In general, 12-heteropoly anions of Si^{+4} and P^{+5} are more readily reduced than those of Ge^{+4}. Furthermore, heteropolymolybdates are more readily reduced than corresponding heteropolytungstates. For example, in the phosphorus acids, the oxidizing power decreases in the order:

1. Dimeric 9-molybdophosphoric(V) acid (strongest oxidant)
2. 12-Molybdophosphoric(V) acid
3. Dimeric α-9-tungstophosphoric(V) acid
4. Dimeric β-9-tungstophosphoric(V) acid
5. 12-Tunstophosphoric(V) acid (weakest oxidant)

Ferrous salts, sulfites, urea, uric acid, hydroquinones, or other mild reducing agents are effective. Moderate reduction proceeds in definite steps corresponding to the addition of 2 or 4 electrons. However, strong reduction, as for example the reduction of 12-heteropolymolybdates with zinc and HCl, disintegrates the complexes completely. In such cases, the reduced products have low molecular weights and give a simple molybdate on reoxidation[23].

Moderate reduction of α-12-molybdosilicic acid with $SnCl_2$ gives two compounds: first a green compound formed by the addition of four electrons; the second, a blue complex formed by the addition of one or more electrons on treatment with more $SnCl_2$. Both reduction products can be reoxidized with HNO_3 in strongly acidic solution to give a quantitative yield of the original alpha acid. On reductions with $SnCl_2$, the beta acid gives only one product, a blue compound.

Salts of the 1:9 arsenic series are more stable than the corresponding phosphorus compounds under most conditions. However, they are slowly converted to colorless salts of lower complexity on long standing with the mother liquor. The 1:9 arsenates are also decomposed by excess sulfuric or hydrochloric acids.

Conversion to Other Complexes

The 9-heteropolymolybdates may represent metastable equilibrium states in some ranges of pH and concentration[4]. Thus dilute free acids prepared by ion exchange contain the 1:9 complexes as virtually the sole anionic species. However, other complexes may develop irreversibly if the temperature is raised above 35 °C. Still other complexes develop irreversibly on long standing. For example, upon standing in some solutions at room temperature, the 9-molybdophosphate complex very gradually converts to the 12-molybdophosphate.

Solutions of heteropolymolybdates appear to contain trace amounts of many of the other possible species in equilibrium (See Table 14). The equilibria are complicated by rate phenomena. However, removal of any one heteropoly species, as by precipitation, eventually leads to complete conversion to that form. Thus, ammonium 9-molybdophosphate in solution will eventually precipitate out as the insoluble ammonium 12-molybdophosphate. Heating accelerates this reaction.

The 9-molybdophosphates may also be converted to 12-molybdophosphates by treatment with acid. In turn, the reverse reaction may be brought about by treatment with base or additional phosphoric acid.

The analogous tungstates exhibit similar behavior.

Some heteropoly acids are easily decomposed in acid solution as has been shown by spectroscopic studies[153]. The order of increasing stability is 12-molybdophosphate, 12-molybdogermanate, 12-molybdosilicate, and 9-molybdophosphate[153].

Preparation

The 9-heteropolymolybdates and 9-heteropolytungstates are prepared at higher temperatures, higher concentrations and under a slightly less acid condition than the 12-anions[7, 23, 27, 28]. The range of conditions necessary for formation is narrower than for other species. Once formed, however, the 1:9 anions remain undecomposed under conditions in which they would not form.

The free acids may be prepared by acidifying solutions of the salt, and extracting with ether. Alternatively, they may be produced from salt solutions by passage through an ion-exchange column. Salts of the arsenic complex may be prepared by simply saturating solutions of arsenates with MoO_3. The 9-molybdophosphoric acid has been prepared by controlled heating of the crystals of the 12-acid at 300 to 350 °C followed by water extraction of the mass[154].

Oxidation-Reduction

The investigation of the oxidation-reduction properties of heteropoly compounds of molybdenum in aqueous and nonaqueous solvents has received increasing attention in recent years. Such knowledge may not only elucidate the redox behavior of such compounds but it could help in the investigation of new preparative procedures for heteropolymolybdates. Such investigation may also be useful in suggesting applications in homogeneous catalysis. Heteropolymolybdates in general are very suitable compounds for such investigation since several of them are known to undergo reversible oxidation-reduction.

Oxidation-Reduction Properties (General)

The 12- and 9-heteropolymolybdates containing nontransition metals as central atoms are relatively strong oxidizing agents[23, 65, 155]. Consequently, they are readily reduced even by mild reducing agents. The corresponding tungstates are not as readily reduced. Heteropolymolybdates with transition metals as central atoms are usually close to tungstates in redox behavior. The reduced anions are dark blue, maintain the same Mo : central atom ratio and show the properties of the parent anions. Many have been isolated as free acids[156, 157] and as salts[23, 156–158]. The reduced acids are readily reoxidized to their original states by bromine water, hydrogen peroxide, or other oxidizing agents formed by the addition of 4 electrons. The compound can be reoxidized, even by air, to the beta acid[65]. The atomic ratio in the reduced anions is 1 Si : 12 Mo.

Electrochemistry

Heteropolymolybdates and their tungsten analogs have been extensively studied polarographically. This literature has been summarized in Refs.[15, 155] and [158–160]. Conflicting results in the literature concerning heteropoly oxidation-reduction reactions apparently arose from failure to take into consideration the hydrolytic instability and reactivity of these compounds with mercury. For example, the polarographic behavior of dilute aqueous solutions of $H_3[PMo_{12}O_{40}]$ has been described in the literature employing the dropping mercury electrode[161]. Further, the polarographic reduction of 12-molybdoarsenic acid, $H_3[AsMo_{12}O_{40}]$, in aqueous solution has been also described[162], although it is now known[15, 36, 114] that the $[AsMo_{12}O_{40}]^{-3}$ anion does not exist in aqueous solution. However, the greater hydrolytic stability and lesser oxidizing power of the corresponding tungstates has permitted their polarographic study in aqueous solutions at the dropping mercury electrode[159]. Consequently, as will be indicated below, the study of the 12-heteropolymolybdates has been carried out employing a platinum electrode in water-dioxane and other mixed solvent media which prevent the hydrolysis of the heteropoly anions[14, 15, 114]. This investigation was carried out by cyclic voltammetry[163, 164] and alternating current polarography[163, 165] which were applied for the first time to elucidate the oxidation-reduction properties of heteropoly compounds[15, 155].

The oxidation-reduction behavior, ascertained by direct current (conventional) polarography, and its dependence on pH in aqueous solution of the $[P_2Mo_{18}O_{62}]^{-6}$, $[As_2Mo_{18}O_{62}]^{-6}$, and $[P_2W_{18}O_{62}]^{-6}$ anions may be found in Refs.[159–161]. The polarographic behavior of the dimeric 9-molybdophosphate anion of $(NH_4)_6[P_2Mo_{18}O_{62}]$ was examined by cyclic voltammetry and alternating current polarography[15, 155] in 1 M H_2SO_4. The values of half-wave potentials thus obtained versus the standard calomel electrode, which are shown in Table 15 and in Fig. 19, are in agreement with those obtained by others[158] for the same compound employing direct current polarography. Addition of more than six electrons to the heteropoly anion in aqueous solution resulted in decomposition of the anion although this effect was not observed in water-dioxane solutions[15, 155]. The reversibility of the oxidation reduction steps for $H_3[PMo_{12}O_{40}]$ in water dioxane solution as ascertained

Table 15. Polarographic data in 1 M H_2SO_4 at 25 °C for $[P_2Mo_{18}O_{62}]^{-6}$ anion

$E_{1/2}$ cathodic (volt)	$E_{1/2}$ anodic (volt)	Number of electrons
+0.47	+0.50	2
+0.35	+0.38	2
+0.175	+0.22	2
−0.10	irreversible	–

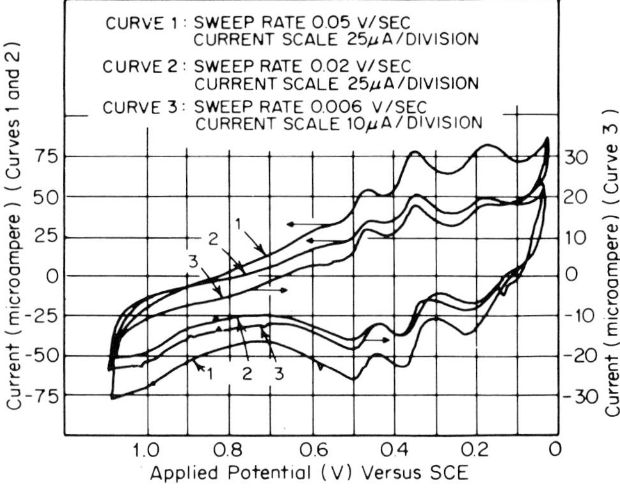

Fig. 19. Cyclic voltammogram of 2 x 10^{-4} M $(NH_4)_6[P_2Mo_{18}O_{62}]$ in 1 M H_2SO_4 at 25 C

from cyclic voltammetry and shown in Fig. 20 was studied at frequencies from 0.01 to 10 Hz. The general shape of the voltammogram indicates some reversibility of the first three waves at the sweep rates studied. The values of the half-wave potentials versus the standard calomel electrode obtained for this compound using a rotating platinum electrode were +3.6, +0.22, −0.01, and −0.15 V[15, 155]. Although the polarograms of $H_4[PMo_{11}VO_{40}]$ and $H_5[PMo_{10}V_2O_{40}]$ in water solution show that these acids are considerably decomposed, the data in water dioxane solutions[15, 155] show that dioxane stabilizes the anion structures toward hydrolysis. Four stepwise reductions obtained for these acids appear to be fairly reversible.

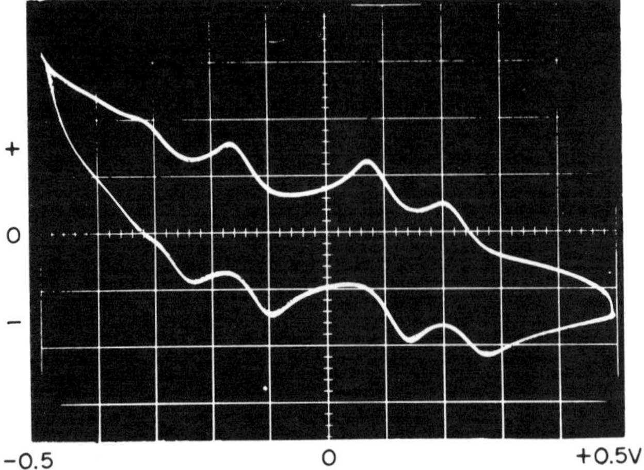

Fig. 20. Cyclic voltammogram of 1×10^{-3} M $H_3[PMo_{12}O_{40}]$ in 1 N H_2SO_4 (1:1 water-dioxane).
Y axis: current (200 μa/div)
X axis: applied potential (0.1 v/div versus SCE
Frequency: 10 Hz
Sweep rate: 20 v/sec

Few studies have been made of the oxidation-reduction properties of heteropolymolybdate anions containing transition metals as central atoms. No "heteropoly blue" involving reduced species of these anions is known. Only a brief report has appeared in the literature[166)] stating that some of these anions are reduced at the dropping mercury electrode, but no definite conclusions can be drawn from these data. The polarographic behavior of the anions $[TeMo_6O_{24}]^{-6}$, $[CrMo_6O_{24}H_6]^{-3}$, $[AlMo_6O_{24}H_6]^{-3}$, $[FeMo_6O_{24}H_6]^{-3}$, $[IMo_6O_{24}]^{-5}$ and of the anion $[Co_2Mo_{10}O_{38}H_4]^{-6}$ has been examined by d.c. and a.c. polarography and by cyclic voltammetry[15, 155)]. All of these anions showed no reduction waves at positive potentials with respect to the SCE, using the platinum electrode. This fact is consistent with the lower oxidizing power of these species compared with the 12-heteropolymolybdates already discussed. All polarographic data on compounds with transition metals as central atoms were taken with the dropping mercury electrode. A cyclic voltammogram of the $[TeMo_6O_{24}]^{-6}$ anion in 1 M NaAc (pH = 3.4) is shown in Fig. 21. It is estimated that the oxidation-reduction step invovles one electron. However, the other six heteropolymolybdates studied exhibited different behavior. For example, the cyclic voltammogram (Fig. 22) of the 6-molybdochromate anion is typical of a system[163)] which undergoes a rapid electron transfer followed by a slow chemical reaction. That is, the anodic peaks are considerably smaller than the cathodic ones (Fig. 22) suggesting that the reduced species undergo decomposition. The cyclic experiments for the other 6-heteropolymolybdates of Al^{+3}, Co^{+3}, Fe^{+3}, I^{+7} and for the anion $[Co_2Mo_{10}O_{38}H_4]^{-6}$ gave voltammograms similar to that of the chromium complex, suggesting that the reduced species produced are unstable in solution, and isolation of any heteropoly blues is thus precluded.

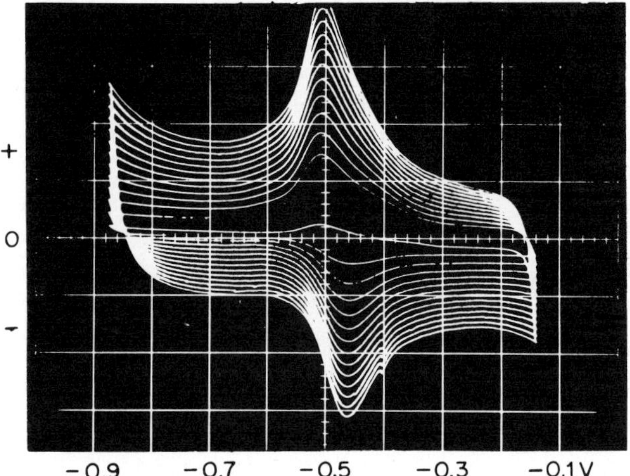

Fig. 21. Cyclic voltammogram of 3 x 10^{-4} M $Na_6[TeMo_6O_{24}]$ in 1 M NaAc-HAc (pH = 3.4).
Y axis: current (5 μa/div
X axis: applied potential (0.1 v/div) versus SCE
Sweep rate (outer cycle): 3.84 v/sec (inner cycles lower sweep rates)

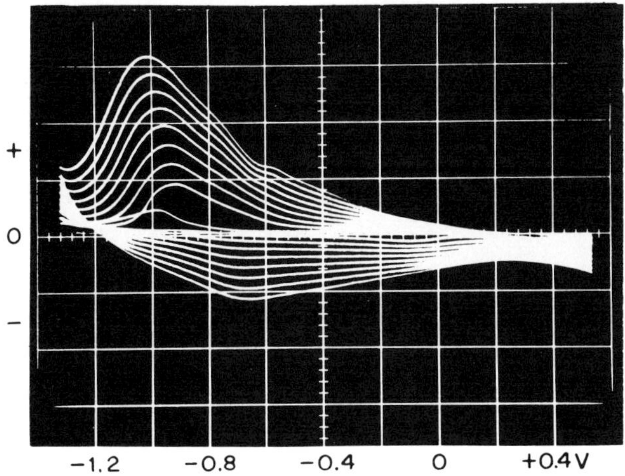

Fig. 22. Cyclic voltammogram of 1 x 10^{-3} M $(NH_4)_3[CrMo_6O_{24}H_6]$ in 1 M $NaClO_4$.
Y axis: current (200 μa/div
X axis: applied potential (0.2 v/div) versus SCE
Sweep rate (outer cycle): 2 v/sec (inner cycles lower sweep rates)

Heteropolymolybdate Blue Species

The favorable reduction potentials of heteropolymolybdates have allowed several workers to isolate heteropoly molybdate blues without difficulties of rapid reoxidation[23, 156, 157, 159, 160, 167–177]. Most of the work has been restricted, however, to the reduction of the 12-molybdophosphate and 12-molybdosilicate anions in aqueous solution. In an early investigation of the reduction of heteropoly 2 : 18 molybdates[23], treatment of acidic solutions of ammonium 18-molybdodiphosphate, $(NH_4)_6[P_2Mo_{18}O_{62}]$, with various reducing agents enabled the isolation of samples of three heteropoly blues. These materials, characterized by ignition and by titration with potassium permanganate, were formulated as the two, four and six-electron reduction products of the $[P_2Mo_{18}O_{62}]^{-6}$ anion. Recent polarographic investigation of the anions $[P_2Mo_{18}O_{62}]^{-6}$ and $[As_2Mo_{18}O_{62}]^{-6}$ has shown[158] that each undergo three successive two-electron reductions in acidic solutions. No intermediate reduction steps were observed. The reduced heteropoly molybdates are weaker acids than the corresponding tungstates and disproportionate rapidly in neutral and alkaline solution. Crystalline samples of ammonium salts and solutions of the free acids of the three reduced molybdophosphates were characterized.

The thermal decomposition of alpha-12-molybdosilicic acid, $H_4[SiMo_{12}O_{40}]$ and its two and four electron reduction products, $H_6[SiMo_{12}O_{40}]$ and $H_8[SiMo_{12}O_{40}]$ has been studied in detail[173–177]. The reduced products decompose at higher temperatures than the oxidized species, the oxides of molybdenum that result from such decomposition have surface areas $\sim 10\ m^2/g$. The reduced product yield Mo_4O_{11} upon decomposition.

The electronic spectra of the one-to-six-electron blue of 18-metallodiphosphates have been studied[178]. The optical spectra of the blues obtained by reducing $[P_2Mo_{18}O_{62}]^{-6}$ show that the added electrons are trapped on the molybdenum atoms. ESR measurements on reduced heteropolymolybdates derived from $[SiMo_{12}O_{40}]^{-4}$ show similar type of behavior even when they contain as many as six extra electrons (or Mo^{+5} atoms)[179]. The formation of reduced heteropolymolybdates has been rationalized with the availability of single unshared terminal oxygen atoms[180].

Other Heteropoly Compounds

The heteropoly anions that do not belong to any of the classifications just discussed are presented below.

Decamolybdodicobaltate(III) Anion

When solutions containing cobaltous and paramolybdate ions are boiled in the presence of an oxidizing agent such as bromine or hydrogen peroxide[7, 8] a blue-green complex is formed of composition $(NH_4)_3[CoMo_6O_{24}H_6] \cdot 7\ H_2O$. A second salt, emerald green in color, which forms during this reaction was formulated in the early literature as $(NH_4)_6[Co_2Mo_{10}O_{36}] \cdot 10\ H_2O$. When persulfate ion is used, only the 6-molybdocobaltate(III) anion is formed. Later studies[150] have shown that when hydrogen peroxide is used as the oxidizing agent in the presence of charcoal, only the $[Co_2Mo_{10}O_{38}H_4]^{-6}$ anion forms. The 6-molybdocobaltate(III) anion is decomposed by the charcoal[150]. The 1 : 5 complex was shown to be dimeric in solution and to have a charge of minus six[150]. A crystal structure determination has shown the proper formula to be $(NH_4)_6[Co_2Mo_{10}O_{38}H_4] \cdot 7\ H_2O$[181]. The anion is dissymmetric (point group D_2) and has been resolved into optical isomers[182]. It is a totally inorganic compound that has been thus resolved. The structure of the decamolybdodicobaltate(III) anion is shown in Fig. 23.

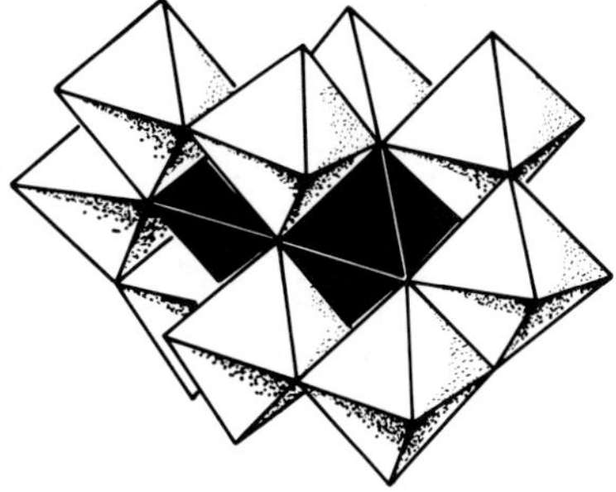

Fig. 23. Structure of the $[Co_2Mo_{10}O_{38}H_4]^{-6}$ anion. The dark octahedra represent $Co^{+3}O_6$ units

12-Molybdotetraarsenate(V) Anion

The crystal structure of ammonium 12-molybdotetraarsenate(V) tetrahydrate has been determined and shown[183, 273] to have the structure depicted in Fig. 24. The most prominent feature of this polyanion is its large cavity at the center surrounded by the eighteen oxygen atoms. The anion is made up of an $Mo_{12}O_{46}^{-20}$ moiety by adding to it four AsO_4 tetrahedra as shown in Fig. 24.

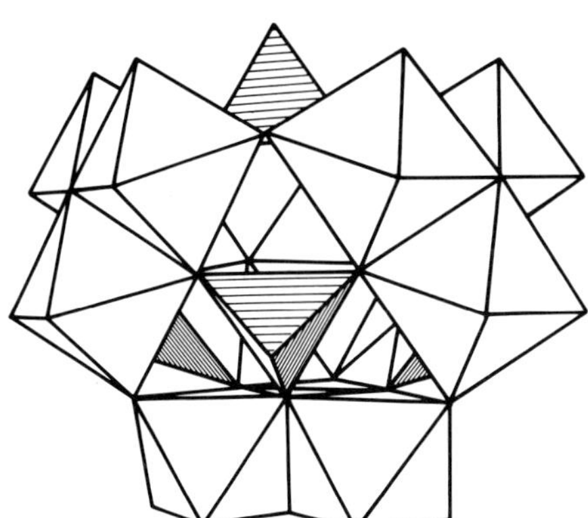

Fig. 24. Model of the $[H_4As_4Mo_{12}O_{50}]^{4-}$ anion constructed by idealized MoO_6 octahedra and AsO_4 tetrahedra

Tetramolybdo Complexes of Dialkyl- and Diarylarsinates

Recently the synthesis and structure of novel heteropoly complexes has been reported that contain covalently attached groups[184]. The complexes have the general formula

$[R_2AsMo_4O_{14}(OH)]^{-2}$ with R = CH_3, C_2H_5 and C_6H_5. The single proton is localized and nontitratable. The heteropoly complexes are readily prepared from stoichiometric quantities of sodium molybdate and the appropriate arsinic acid at pH 4–5. The anions formed are stable in the pH range 2–6. The structure of the anion, shown in Fig. 25, consists of a ring of four alternately face- and edge-shared MoO_6 octahedra capped by the $(CH_3)_2AsO_2$ tetrahedron as shown in Fig. 25. The structure represents

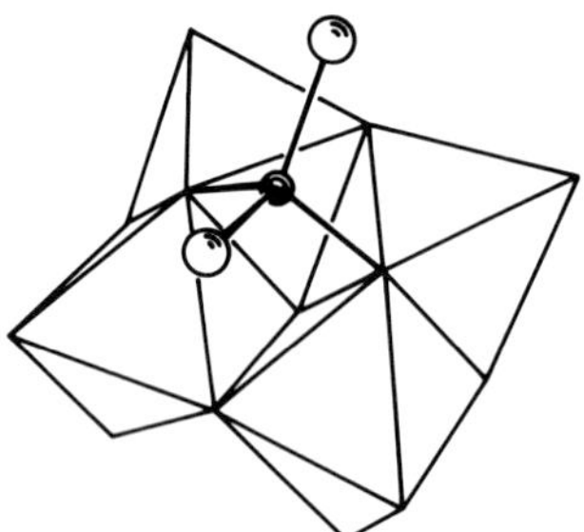

Fig. 25. Structure of $[(CH_3)_2AsMoO_{14}OH]^{2-}$: O, CH_3; •, As; octahedra represent MoO_6 groups

only the second example of a heteropoly complex containing face-shared octahedra, the other being the $[CeMo_{12}O_{42}]^{-8}$ anion shown in Fig. 14.

Three polymolybdosulfate(IV) ions, $[S_2Mo_8O_{31}]^{-6}$, $[S_2Mo_5O_{21}]^{-4}$ and $[S_3Mo_{10}O_{40}]^{-8}$ have been reported in the literature, but the existence of $[S_3Mo_{10}O_{40}]^{-8}$ has been regarded as doubtful. Recent preparative, X-ray and IR work has confirmed the first two polyanions as ammonium salts, $(NH_4)_6[S_2Mo_8O_{31}] \cdot 5\ H_2O$ and $(NH_4)_4[S_2Mo_5O_{21}] \cdot 3\ H_2O$[274] The crystal structure of the ammonium pentamolybdodisulfate(IV) has been determined[274].

The solid is neither an inclusion nor a clathrate compound of sulfur dioxide gas but contains a discrete polyanion consisting of two SO_3 trigonal pyramids and five

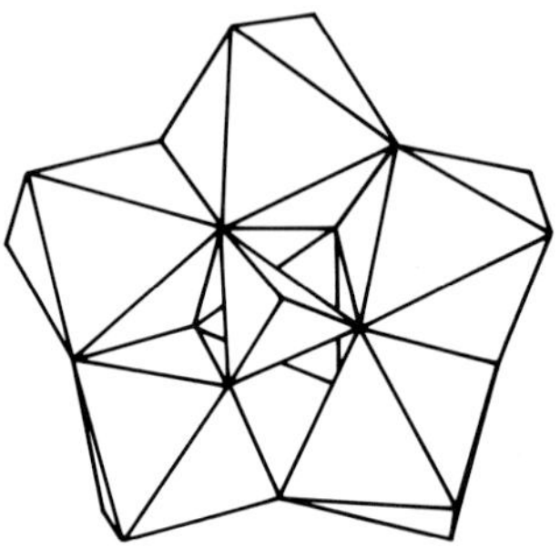

Fig. 26. The structure of $[S_2Mo_5O_{24}]^{4-}$ anion. Sulfur atoms occupy the apexes of the central trigonal pyramids. Five molybdenum atoms lie nearly on a plane to form a pentagon

MoO_6 octahedra as shown in Fig. 26. The following reactions depict the formation of the polyanion in solution:

$$5\ Mo_7O_{24}^{-6} + 14\ SO_2 + 2\ H^+ \rightleftharpoons 7\ [S_2Mo_5O_{21}]^{-4} + H_2O$$

$$5\ MoO_4^{-2} + 2\ SO_2 + 6\ H^+ \rightleftharpoons [S_2Mo_5O_{21}]^{-4} + 3\ H_2O$$

Mixed Heteropoly Complexes

Molybdenum and tungsten form mixed heteropoly anions in which some of the MoO_6 octahedra surrounding the central atom have been replaced by WO_6. Definite compositions may be obtained, but the exact placement of the MoO_6 and WO_6 octahedra is not known. Numerous mixed heteropoly complexes have been reported in the literature and references to these are to be found in the references cited. Only a brief account of these is given here. The 12-heteropolymolybdotungstates with phosphorus or silicon as central atoms have been reported in the literature[185]. These are obtained from solutions of sodium phosphate or sodium silicate and varying quantities of sodium paratungstate and sodium paramolybdate and acidifying the resulting solutions. The acids thus obtained correspond to compositions $H_3[PMo_nW_mO_{40}]$, where n + m = 12.

The properties of these mixed compounds are usually intermediate between those of the corresponding heteropolymolybdates and heteropolytungstates. For example, the sensitivity to reduction of the mixed heteropolytungstomolybdates increases with molybdenum content. Thus colors formed by precipitation of basic dyes with 12-molybdophosphoric acid tend to darken and lose brilliance under exposure to light. The analogous 12-tunstophosphate pigments fade. The mixed tungstomolybdophosphoric acid (PTMA) gives pigments of stable intermediate properties[186].

Mixed heteropolymolybdates and tungstates are present in the 6-heteropoly series. The synthesis of the ammonium salts of the 6-heteropoly anions with nickel as the central atom of the general formula $(NH_4)_4[NiMo_{6-n}W_nO_{24}H_6]$ has been reported[187]. All compounds thus obtained are light blue. The preparation of the free acids has also been described[187].

Another class of 12-heteropoly anions involves the replacement of molybdenum by pentavalent vanadium. The molybdovanadophosphoric acids, $H_4[PMo_{11}VO_{40}] \cdot 32\,H_2O$, $H_5[PMo_{10}V_2O_{40}] \cdot 32\,H_2O$, and $H_6[PMo_9V_3O_{40}] \cdot$ $\cdot\, 34\,H_2O$ were prepared as orange crystalline solids and characterized[47]. The heteropoly anions were found to have unusual hydrolytic stability. Details of properties of these compounds are given in the literature[47]. The preparation of the salts $Na_4[PMo_{11}VO_{40}] \cdot 8\,H_2O$, $(NH_4)_3H[PMo_{11}VO_{40}] \cdot 7.5\,H_2O$, and $(NH_4)_3H_2[PMo_{10}V_2O_{40}] \cdot 7.5\,H_2O$ has also been described[47].

The preparation of the free 10-tungsto-2-vanadophosphoric acid, $H_5[PW_{10}V_2O_{40}]$, has been described, as well as that of the 8-tungsto-4-vanadophosphoric acid[188].

Niobium, like vanadium, can replace molybdenum in the heteropoly anions. The yellow 10-molybdo-2-niobophosphate anion has been prepared[104]. The salts $(C_5H_5NH)_4H[PMo_{10}Nb_2O_{40}]$ and $[(CH_3)_4N]_4H[PMo_{10}Nb_2O_{40}]$ have been prepared and characterized. The free acid could not be isolated since it is unstable even in solution[104].

A tungstovanadoselenous acid of composition $H_6[SeMo_{10}V_2O_{40}] \cdot xH_2O$ has been prepared by ether extraction and partly characterized[189].

In the Keggin structure of the 12-heteropolymolybdates and 12-heteropolytungstates, it is possible to substitute atoms other than vanadium or niobium for tungsten or molybdenum.

Heteropoly anions of general formula $[XZW_{11}O_{40}H_2]^{-n}$ have been prepared[190] where X = Si, Ge, or P, and Z = Co(II), Co(III), or Ni(II). In these series, X is the heteroatom that occupies the tetrahedral position within the Keggin structure, and Z acts as a perihedral atom along with tungsten or molybdenum. Table 16 identifies the heteropoly anions that have been obtained[190].

Table 16. Summary data on the heteropoly anions

Anion $[XZA_{11}O_{40}H_2]^{-n}$	I	II	III	IV	V	VI	VII
Central heteroatom X	Si	Ge	Ge	Si	Ge	P	Si
Outer heteroatom Z	Co(III)	Co(II)	Co(III)	Ni(II)	Ni(II)	Ni(II)	Co(II)
Addendum atoms A	W	W	W	W	W	W	Mo
Charge	−5	−6	−5	−6	−6	−6	−6

Similar substitution has also been extended in the dimeric 9-heteropolytungstates typified by the $[P_2W_{18}O_{62}]^{-6}$ anion[191, 192]. The preparation of eleven heteropoly-tungstate anions, $[X_2ZW_{17}O_{62}H_2]^{-n}$ X = P or As; Z = Mn(II), Mn(III); Co(II), Co(III); Ni(II); or Zn(II), has been described recently[191]. Another work[192] describes similar compounds with Z = Mn(II), Fe(III), Co(II), Ni(II), Cu(II), Zn(II), Cr(III), and Ti(IV). Detailed description of these compounds are to be found in the references cited. The general preparative procedure involves the formation of a 17-tungstodi-phosphate anion followed by the addition of the desired metal salt[192].

A large number of other heteropolymolybdates or heteropolytungstates has been reported in the literature[8, 24]. These are frequently single compounds and not members of any series. In many cases, the compounds are probably definite double salts or acid salts. Further work would be necessary to clarify the nature of these systems.

Spectra of Heteropoly Compounds

Infrared

Infrared spectra of some heteropoly compounds have been examined[47, 100, 193]. In general, the Mo—O and W—O bonds absorb in the 1000 to 800 cm^{-1} region. In the 12-heteropolytungstates, the Si—O and P—O absorptions of the central tetrahedron occur at 1000 and 530 cm^{-1} for the Si—O and at 1075 and 520 cm^{-1} for P—O. For Zn—O, bands are at 449 and 248 cm^{-1}. An unusually high value of a band at 1170 cm^{-1} in several heteropolytungstates has been attributed to W—O[100]. The highest absorption for the W—O bond in WO_4^{-2} occurs at 928 cm^{-1} [194]. It may therefore be concluded that the 1170 cm^{-1} band is associated with the W—O bond involving the nonbridged peripheral oxygen atoms. This very short W—O distance present in heteropolytungstates is consistent with X-ray W—O distance values.

Lattice water present in heteropoly compounds absorbs in the region 3550 to 3200 cm^{-1} and also around 1600 cm^{-1}.

The infrared spectra of the compounds $H_3[PMo_{12}O_{40}] \cdot 15\ H_2O$, $H_4[PMo_{11}VO_{40}] \cdot 35\ H_2O$, $H_5[PMo_{10}V_2O_{40}] \cdot 32\ H_2O$, and $H_6[PMo_9V_3O_{40}] \cdot$ $\cdot\ 33\ H_2O$ were obtained[47], but in these the strong obsorption band found at 1170 cm^{-1} in the 12-heteropolytungstates and attributed to W—O bonding[100] was not found.

Visible and Ultraviolet Spectra

The absorption spectra of heteropoly anions in the visible range are of two kinds:

(a) those involving d-d transitions present in heteropoly anions containing transition elements as central atoms, and

(b) heteropoly anions with central atoms other than transiton elements having charge transfer adsorptions with motor extinction coefficients of the order of 10^4 to 10^5 which may be apparent in the visible spectra as found in the reduced 12-molybdosilicate anion.

For example, the $[P_2Mo_{18}O_{62}]^{-6}$ anion is yellow whereas $[P_2W_{12}O_{62}]^{-6}$ is very light yellow. In general, for the same type of complex, molybdates absorb at longer wavelengths than the corresponding tungstates[127]. Information of this type leads to an explanation of the coordination about the central atom and distinction between various structural isomers such as the α and β forms of the $[P_2W_{18}O_{62}]^{-6}$ anion[195].

Spectra of heteropolymolybdates are also to be found in references cited dealing with the individual compounds. Recently, electron spin resonance spectra of molybdovanadophosphoric acids have also been reported[196].

Physical Measurements

Studies of the solution properties of heteropoly acids have been somewhat spares despite the general interest in these compounds for many years. Deterents to such studies have been primarily the instability of the compounds and the uncertainty concerning their composition. Conductivity and pH measurements on the heteropoly acids $H_4[PMo_{11}VO_{40}]$ and $H_5[PMo_{10}V_2O_{40}]$ in aqueous solutions and mixed solvents has already been discussed. The acids are strong 1—4 and 1—5 electrolytes, respectively. Activity coefficients of ammonium 6-heteropolymolybdates have been reported and shown these to be 1:3 electrolytes[197].

The diffusion coefficients of the 12-tungstosilicate anion $[SiW_{12}O_{40}]^{-4}$ and its first two reduced forms $[SiW_{12}O_{40}]^{-6}$ and $[SiW_{12}O_{40}]^{-8}$ have been determined[198] and found to be of the order $4.1 \times 10^{-6}\ cm^2/sec$. Diffusion coefficients for $[SiMo_{12}O_{40}]^{-4}$ and $[SiW_{12}O_{40}]^{-4}$ have also been reported by others[70, 199]. The diffusion coefficients of $[CoMo_6O_{24}H_6]^{-3}$ and $[CrMo_6O_{24}H_6]^{-3}$ were found to be in the order of $7.4 \times 10^{-6}\ cm^2/sec.$[70].

Other studies on heteropoly anions include electron paramagnetic resonance on the reduced forms of heteropoly compounds[196, 198] and thermal and *in situ* X-ray studies of some heteropoly compounds[200]. No thermodynamic data are available on heteropoly compounds.

Uses

A considerable amount of literature has appeared which describes the use of heteropolymolybdates and their tungsten analogs in various applications. Of these, the main applications have centered on catalysis. Of particular importance to the catalytic behavior of heteropoly compounds is the solubility and solvolytic behavior in both aqueous and organic media and the thermal stability and oxidation-reduction behavior.

Catalysis

The various applications of heteropoly compounds in catalytic processes up to 1950 have been compiled in Ref.[32] and for subsequent years in molybdenum catalyst bibliographies that are issued every two years[201–204]. The examples listed below will serve to illustrate the diversity of applications in the area in question.

12-Molybdophosphoric and 12-molybdosilicic acids and several of their metal salts have been used as catalysts in hydrodesulfurization[205], epoxidation of olefins[206], alkylation[207, 208], preparation of saturated carbonyl compounds[209], and in the direct oxidation of benzene to phenol[210]. Ammonium 12-molybdocerate promoted with tellurium has been described as a catalyst in the production of acrylonitrile from propylene, air and ammonia[211]. Heteropoly compounds have also been used as catalysts for the vapor phase partial oxidation of naphthalene[212] and the vapor phase hydration of ethylene[213]. The 12-tungstophosphoric and 12-tungstosilicic acids have been reported as the most effective catalysts for the dehydration of castor oil to unsaturated oils[214]. 12-Molybdophosphoric acid catalyzes the conversion of unsaturated hydrocarbons to carbonyl compounds[215] and the oxidation of olefins in the presence of tellurium or a tellurium compound to give α, β-unsaturated carboxylic acids[216]. Heteropoly acids have been used in the polymerization of olefins[217–219] and the spent catalyst could be regenerated[219]. The 12-tungstophosphoric acid catalyzes the hydroxylation of alkyl alcohol to glycerol using hydrogen peroxide[220, 221]. The 12-molybdophosphoric and 12-molybdosilicic acids have been reported to catalyze the hydrogenation of aromatic hydrocarbons[222]. Silicon carbide impregnated with 12-molybdophosphoric acid catalyzes the conversion of unsaturated aldehydes into the corresponding carboxylic acid[223]. Methacrolein, acrylic, acid, and methacrylic acid are prepared by the catalyzed oxidation of propylene or isobutylene in the presence of ammonium 9-molybdomanganate(IV) and arsenic oxide[224].

The catalytic oxidation in the presence of various heteropoly compounds of lower olefins to unsaturated aldehydes and subsequent conversion into unsaturated nitriles are described in Ref.[225–231]. Copper phthalocyanine is produced in 92% yield from phthalic anhydride in the presence of 12-molybdophosphoric acid[232].

Flame Retardants

Molybdenum compounds can act as flame retardants and smoke suppressants in textiles and in plastics[233, 275]. Heteropoly compounds have been found to act as flame retardants for wood[234–236] and for textiles[233]. Flame-proofing action on polymethylmethacrylate has been reported[237].

Corrosion Inhibition

The application of heteropolymolybdates as corrosion inhibitors in aqueous solution must be approached with caution since some heteropoly compounds undergo hydrolytic degradation in very dilute solutions and are not stable above pH 3.5. For example, the reported corrosion inhibition of sodium 12-molybdophosphate was shown to be due to sodium molybdate and sodium phosphate rather than to the heteropoly anion[238]. However, heteropoly compounds have been used as conversion (reaction) coatings on steel[239] and aluminum[240] and as organic coatings on steel[241, 242] with anticorrosion properties. Their use on the corrosion inhibition of steel has been also reported[243].

Miscellaneous

Certain heteropolymolybdates and heteropolytungstates, notably the 12-heteropolymolybdates and 12-heteropolytungstates, are produced in large quantities by the color industry as precipitants for basic dyes, with which they form lakes or toners[186]. The formation of basic dye-heteropoly acid complexes for printing anionic synthetic fibers has also been reported[244]. In biological and analytical chemistry, 12-heteropolymolybdates and 12-heteropolytungstates of phosphorus and silicon have been widely used as reagents[23, 245]. Phosphorus, silicon and arsenic can be determined in the presence of one another by formation of the corresponding 12-heteropolymolybdates and subsequent selective extraction of the acids[124, 246]. Phosphorus is determined colorimetrically after formation of 12-molybdophosphate anion or molybdovanadophosphate[247], and arsenic is determined similarly as the tungstovanadoarsenic acid[247]. Trace amounts of colloidal platinum can be determined by its catalytic action on the reduction of 12-molybdophosphoric acid by formic acid to form molybdenum blue[248]. Colorimetric determinations employing 12-molybdophosphoric acid include determination of catechol and its methylhomologue[249], penicillin in procaine penicillin G[250], and the determination of dihydroxypropyl-theophylline by means of 12-molybdophosphoric acid[251]. This acid has been employed in developing chromatograms for thiamine[252] and methyl testosterone[253].

12-Molybdosilic acid has been employed in the isolation of alkaloids from biological fluids[254] and as a reagent for choline[255]. The sensitivity of alkaloids and certain nitrogen-containing bases toward 12-tungstophosphoric acid has been utilized for the turbidimetric determinations of small concentrations of atropine, atropine sulfate, and caffeine[256]. Benzalkonium chloride is gravimetrically determined after precipitation with 12-tungstophosphoric acid[257]. 12-Molybdosilicic acid and the phosphorus analogue were used as spray reagents for separation by thin-layer chromatography[258]. Ammonium 9-molybdophosphate has been shown to be a sensitive reagent for the determination of phenols and pyrroles[259].

Cigarette filters impregnated with 12-tungstophosphoric acid were shown to be efficient in removing nicotine (up to 50%) from smoke[260]. Potassium, rubidium and cesium have been separated from fission product solutions after coprecipitation with 12-tungstophosphoric acid and subsequent separation by ion exchange[261]. Other applications include the use of heteropoly compounds as additives in lubricating greases having extreme-pressure characteristics and resistance to water[262] and the development of an inorganic phototropic system for flashblindness protection[263]. Photochromic layers containing 12-molybdosilicic acid and phenylglycolic acid are stabilized by the use of the molybdenum compound[264]. Heteropolymolybdates have also been used in the preparation of planographic printing plates[265].

References

1) Evans, Jr., H. T.: Perspectives Struct. Chem. *4,* 1 (1971); Weakly, T. J. R.: Structure and Bonding *18,* 131 (1974)
2) Kepert, D. L.: Comprehensive inorganic chemistry. (A. F. Trofman, *et al.*) Oxford: Pergamon Press, 1973, Vol. 4, pp. 607
3) Kazanskii, L. P., Turchenkova, E. A., Spitsyn, V. I.: Uspekii Khimii *43,* 1137 (1974)
4) Souchay, P.: Polyanions et polycations. Paris: Gauthier Villars Editeur 1963
5) Tsigdinos, G. A.: Ind. Eng. Chem., Prod. Res. Develop. *13,* 267 (1974)
6) Tsigdinos, G. A.: Heteropoly compounds of molybdenum and tungsten. Climax Molybdenum Company Bulletin Cdb-12a (Revised), November, 1969
7) Tsigdinos, G. A.: Heteropoly-Verbindungen (Methodicum Chimicum), Niedenzu, K., Zimmer, H. (eds.), Vol. 8, Chapter 32. Stuttgart: Georg Thieme Verlag, 1974
8) Gmelins: Handbuch der Anorganischen Chemie. System Number 53 (Molybdenum). Berlin: Verlag Chemie 1935
9) Struve, H.: J. Prakt. Chem. *55,* 888 (1854)
10) Marignac, C.: J. Prakt. Chem. *77* (2), 417 (1862)
11) Rosenheim, A.: Handbuch der Anorganischen Chemie. Abegg, R. (ed.), Vol. 4, Part 1, ii, pp. 977–1065. Leipzig: S. Hirzel 1921
12) Hegedus, A., Dvorsky, M.: Magr. Tud. Akad. Kem. Tud. Oszt. Kozlemen. *11,* 327 (1959)
13) Nomenclature of inorganic chemistry, in: J. Am. Chem. Soc. *82,* 5523 (1960)
14) Souchay, P., Contant, R.: C. R. Acad. Sci., Paris, Ser. C. *265*, 723 (1967)
15) Tsigdinos, G. A., Hallada, C. J.: J. Less-Common Met. *36,* 79 (1974)
16) Massart, R., Souchay, P.: C. R. Acad. Sci., Paris *257,* 1297 (1963)
17) Massart, R., Herve, G.: Rev. Chim. Minerale *5,* 501 (1968)
18) Rabette, P.: Thesis, University of Paris VI, 1973, p. 2
19) Inorganic synthesis, Vol. I. New York: McGraw-Hill Book Company, Inc. 1939, pp. 127–133
20) Drechsel, E.: Ber. *20,* 1452 (1887)
21) Baker, L. C. W., Loev, B., McCutcheon, T. P.: J. Am. Chem. Soc. *72,* 2374 (1950)
22) Meiklejohn, P. T., Pope, M. T., Prados, R. A.: J. Am. Chem. Soc. *96,* 6780 (1974)
23) Wu, H.: J. Biol. Chem. *43,* 189 (1920)
24) Gmelins: Handbuch der Anorganischen Chemie. System Number 54 (Tungsten). Berlin: Verlag Chemie 1933
25) Mellor, J. W.: A comprehensive treatise on inorganic and theoretical chemistry. Vol. XI (Molybdenum and Tungsten). London: Longmans Green and Co. 1931
26) Jolly, W. L.: The synthesis and characterization of inorganic compounds. Englewood Cliffs, N. J.: Prentice-Hall, Inc. 1970, p. 460
27) Brauer, G. (ed.): Handbook of preparative inorganic chemistry. 2nd edit., Vol. 2. New York: Academic Press, Inc. 1965, pp. 1698–1740 (English Translation)

28) Malaprade, L.: Heteropolyacides derives des anhydrides molybdique et tungstique. Pascal, P. (ed.): Nouveau traite de chimie minerale, Vol. 14. Paris: Masson et Cie 1959, pp. 903–981
29) Souchay, P., Faucherre, J.: Bull. Soc. Chim. Fr. *18,* 355 (1951)
30) Baker, M. C., Lyons, P. A., Singer, S. J.: J. Phys. Chem. *59,* 1074 (1955)
31) Linz, A.: Ind. Eng. Chem., Anal. Ed. *15,* 459 (1942)
32) Killefer, D. H., Linz, A.: Molybdenum compounds. New York: Interscience Publishers 1952, pp. 87–94
33) Massart, R., Souchay, P.: Compt. Rend. *257,* 1297 (1963)
34) Massart, R.: Ann. Chim. (Paris) *3,* 507 (1968)
35) Tsigdinos, G. A., Hallada, C. J.: Synthesis and electrochemical properties of heteropolymolybdates, Presented at First Conference of Molybdenum Chemistry, University of Reading, September 1973
36) Strickland, J. D. H.: J. Am. Chem. Soc. *74,* 862 (1952)
37) Barbieri, G. A.: Atti Accad. Naz. Lincei, Rend. Cl. Sci. Fis. Mat. Nat. Ser. 5, *21,* Part I, 781 (1913)
38) Torchenkova, E. A., Baidala, P., Smurova, V. S., Spitsyn, V. I.: Dokl. Akad. Nauk SSSR *199,* 120 (1971)
39) Baidala, P., Smurova, V. S., Torchenkova, E. A., Spitsyn, V. I.: Dokl. Akad. Nauk SSSR *197,* 339 (1971)
40) Tsigdinos, G. A.: Ph. D. Thesis, Boston University, 1961, p. 106
41) d'Amour, H., Allmann, R.: Naturwissenschaften *61,* 31 (1974)
42) Malik, S. A.: J. Inorg. Nucl. Chem. *32,* 2425 (1970)
43) Fournier, M., Massart, R.: C. R. Acad. Sci. Ser. C, *276,* 1517 (1973)
44) Tourne, C., Tourne, G., Malik, S. A., Weakley, T. J. R.: J. Inorg. Nucl. Chem. *32,* 3875 (1970)
45) Leyrie, M., Fournier, M., Massart, R.: C. R. Acad. Sci. Ser. C, *273,* 1569 (1971)
46) Petit, M., Massart, R.: C. R. Acad. Sci., Paris Ser. C, *268,* 1860 (1969)
47) Tsigdinos, G. A., Hallada, C. J.: Inorg. Chem. *7,* 437 (1968)
48) Strandberg, R.: Acta Chem. Scand. *27,* 1004 (1973)
49) Hedman, B.: Acta Chem. Scand. *27,* 3335 (1973)
50) Bhattacharya, G. C., Roy, S. K.: J. Indian Chem. Soc. *50,* 359 (1973)
51) Pauling, L.: The nature of the chemical bond, 3rd Edit. Ithica, New York: Cornell University Press 1960, p. 514
52) Linqvist, I.: Arkiv Kemi *2,* 349 (1950)
53) Shimao, E.: Bull. Chem. Soc. Jap. *40,* 1609 (1967)
54) Evans, H. T., Jr.: J. Am. Chem. Soc. *90,* 3275 (1968)
55) Evans, H. T., Jr., Gatehouse, B. M., Leverett, P.: J. Chem. Soc., Dalton Trans. 505 (1975)
56) Linett, J. W.: J. Chem. Soc. *1961,* 3796
57) Yannoni, N. F.: Ph. D. Dissertation, Boston University, 1961
58) Perloff, A.: Inorg. Chem. *9,* 2228 (1970)
59) Hückel, W.: Structural chemistry of inorganic compounds. Vol. I. New York and Amsterdam: Elsevier Publishing Co. 1950, pp. 179–213
60) DeA. Santos, J. R.: Dissertation, University of Coimbra, Coimbra, Portugal (1947)
61) DeA. Santos, J. R.: Proc. Roy. Soc. (London) *A150,* 309 (1935)
62) Levy, H. A., Agron, F. A., Danford, M. D.: J. Chem. Phys. *30,* 1486 (1959)
63) Babad-Zakhryapin, A. A., Gorbunov, N. S.: Izv. Akad. Nauk SSR, Otd. Khim. Nauk 1870 (1962)
64) Glemser, O, Holznagel, W., Höltje, W., Schwarzmann, E.: Z. Naturforsch. *20b,* 725 (1965)
65) Strickland, J. D. H.: J. Am. Chem. Soc. *74,* 862, 868, 972 (1952)
66) Massart, R., Fournier, M., Souchay, P.: Compt. Rend. *267,* Ser. C, 1850 (1968)
67) Langer, K.: Z. Anal. Chem. *245,* 139 (1969)
68) Souchay, P.: Bull. Soc. Chim. Fr. *18,* 365 (1951)
69) Yamamura, K., Sasaki, Y.: J. Chem. Soc., Chem. Comm. *1973,* 649
70) Baker, L. C. W., Pope, M. T.: J. Am. Chem. Soc. *82,* 4176 (1960)

71) Signer, R., Gross, H.: Helv. Chim. Acta *17,* 1076 (1934)
72) Kuruscev, T., Sargeson, A. M., West, B. O.: J. Phys. Chem. *61,* 1569 (1957)
73) West, S. F., Audrieth, L. F.: J. Phys. Chem. *59* 1069 (1955)
74) Dawson, B.: Acta Cryst. *6,* 113 (1953)
75) Van R. Smit, J., Jacobs, J. J., Robb, W.: J. Inorg. Nucl. Chem. *12,* 95 (1959)
76) Krtil, J., Krivy, I.: J. Inorg. Nucl. Chem. *25,* 1190 (1963)
77) Choudhouri, D., Mukherjee, S. K.: J. Inorg. Nucl. Chem. *30,* 3091 (1968)
78) Agarwala, U. C.: Ph. D. Dissertation, Boston University, 1960.
79) La Ginestra, A., Cerri, R.: Gazz. Chim. Ital. *95,* 26 (1965)
80) La Ginestra, A., Seta, M.: Ric. Sci. *37,* 287 (1967)
81) Bradley, A. J., Illingsworth, J. W.: Proc. Roy. Soc. (London), *A157,* 113 (1936)
82) Wells, A.: Structural inorganic chemistry, 3rd edit. Oxford: The Clarendon Press 1962, p. 451
83) Wada, T.: Compt. Rend. *259,* 553 (1964)
84) Chuvaev, V. F., Spitsyn, V. I.: Dokl. Akad. Nauk. SSSR *166,* 160 (1966)
85) Lunk, K. I., Spitsyn, V. I.: Dokl. Akad. Nauk SSSR *181,* 1156 (1968)
86) Chuvaev, V. F., Baidola, P., Torchenkova, E. A., Spitsyn, V. I.: Dokl. Akad. Nauk SSSR *196,* 1097 (1971)
87) Gazarov, R. A., Chuvaev, V. F., Spitsyn, V. I.: Russ. J. Inorg. Chem. *19,* 1038 (1974)
88) Chuvaev, V. F., Shinik, G. M., Polotebnoba, N. A., Spitsyn, V. I.: Russ. J. Inorg. Chem. *19,* 633 (1974)
89) Baker, L. C. W., Foster, G., Tan, W., Scholinik, F., McCutcheon, T. P.: J. Am. Chem. Soc. *77,* 2136 (1955)
90) Barbier, A., Malaprade, L.: Compt. Rend. *256,* 168 (1963)
91) Shakova, Z. F., Gavrilova, S. A.: Zh. Neorg. Khim. *3,* 1370 (1958)
92) Dexter, D. D., Silverton, J. V.: J. Am. Chem. Soc. *90,* 3589 (1968)
93) Souchay, P., Tchakirian, A.: Ann. Chim. (Paris) *1,* 249 (1946)
94) Schumb, W. C., Hartford, W. H.: J. Am. Chem. Soc. *56,* 2613 (1934)
95) Keggin, J. F.: Nature, *131,* 908 (1933); Proc. Roy. Soc. (London), *A144,* 75 (1934)
96) Liberti, A., Giombini, G., Cervone, E.: Ric. Sci. *25,* 880 (1955)
97) Liberti, A., Santoro, C.: Ric. Sci. *24,* 2079 (1954)
98) Kraus, O.: Z. Krist. *91,* 402 (1935); *93,* 379 (1936)
99) Illingsworth, J. W., Keggin, J. F.: J. Chem. Soc. 575 (1935)
100) Brown, D. H.: Spectrochim. Acta *19,* 585 (1962)
101) Matijevic, E., Kerker, M.: J. Am. Chem. Soc. *81,* 1307 (1959)
102) Kapustinskii, A. F.: Quart. Rev. *10,* 284 (1956)
103) Moody, G. J., Thomas, J. D. R.: J. Chem. Ed. *42,* 204 (1965)
104) Tsigdinos, G. A.: Climax Molybdenum Company, unpublished results
105) Kourin, V.: Jaderna Energie *9,* (4), 125 (1963)
106) Jean, M.: Ann. Chim. (Paris) *3,* 470 (1948)
107) Matijevic, E., Kerker, M.: J. Am. Chem. Soc. *62,* 1271 (1958)
108) Ripan, R.: Compt. Rend. *227,* 474 (1948)
109) Thiestlethwaite, W. P.: J. Inorg. Nucl. Chem. *19,* 1581 (1967)
110) Goehring, J. B., Kerker, M., Matijevic, E., Tyree, S. Y., Jr.: J. Am. Chem. Soc. *81,* 5280 (1959)
111) Matijevic, E., Kerker, M.: J. Am. Chem. Soc. *81,* 5560 (1959)
112) Keller, J. R., Matijevic, E., Kerker, M.: J. Am. Chem. Soc. *65,* 56 (1961)
113) Hallada, C. J., Tsigdinos, G. A., Hudson, B. S.: J. Phys. Chem. *72,* 4304 (1968)
114) Tsigdinos, G. A., Hallada, C. J.: Inorg. Chem. *9,* 2488 (1970)
115) Johnson, J. S., Kraus, K. A., Scatchard, G.: J. Phys. Chem. *64,* 1867 (1960)
116) Johnson, J. S., Rush, R. M.: J. Phys. Chem. *72,* 360 (1968)
117) Rode, E. Y., Sokolova, M. P., Russ.: J. Inorg. Chem. *8,* 980 (1963)
118) Brown, D. H.: J. Chem. Soc. 3189 (1962)
119) Nikitina, E. A., Buris, E. Y.: J. Inorg. Chem. USSR *3,* 2694 (1958)
120) Brown, D. H.: J. Chem. Soc. *1962,* 3322
121) Brown, D. H.: J. Chem. Soc. *1962,* 4408

122) Newkirk, A. E., Simons, E. L.: Talanta *13,* 1401 (1966)
123) Shimura, Y., Tsuchida, R.: Bull. Chem. Soc. Jap. *30,* 502 (1957)
124) Clabaugh, W. S., Jackson, A.: J. Res. Nat. Bur. Stand. *62,* 201 (1959)
125) Beiles, R. G., Rozmanova, Z. E., Andreeva, O. B.: Russ. J. Inorg. Chem. *14,* 1122 (1969)
126) Fischer, J., Picard, L., Toledano, P.: J. Chem. Soc., Dalton Trans. *1974,* 941
127) Tchakirian, A., Souchay, P.: Ann. Chim. (Paris) *1,* 232 (1946)
128) T'an-Lung, Sudakov, F. P., Shakhova, Z. F.: Zh. Anal. Khim. *19,* 734 (1964)
129) Barbieri, G. A.: Atti Accad. Naz. Lincei, Rend., Cl. Sci. Fis. Mat. Nat. Ser. 5, *23,* Part I, 805 (1914)
130) Baker, L. C. W., Gallagher, G. A., McCutcheon, T. P.: J. Am. Chem. Soc. *75,* 2493 (1953)
131) Shakhova, Z. F., Gavrilova, S. A.: J. Inorg. Chem. USSR *8,* 1370 (1958)
132) Barbieri, G. A.: Atti Accad. Naz. Lincei, Rend., Cl. Sci. Fis. Mat. Nat. Ser. 5, *21,* Part I, 781 (1913)
133) Shakhova, Z. F., Gavrilova, S. A., Zakharova, V. F.: Russ. J. Inorg. Chem. *7,* 904 (1962)
134) Kazanski, L. P.: Thesis, Moscow State University, 1973
135) Waugh, J. L. T., Mair, J. A.: J. Chem. Soc. *1950,* 2372
136) Ganelina, E. S., Matyazh, P. Y.: Zh. Neorg. Khim. *8,* 1891 (1963)
137) Allen, E. R., Benkenkamp, J.: U.S. Atomic Energy Commission, Reports NYO-3549 and NYO-3550 (1954)
138) Waugh, J. L. T., Shoemaker, D. P., Pauling, L. C.: Acta Cryst. *7,* 438 (1954)
139) Liquori, A., Bertinotti, F.: Atti Accad. Naz. Lincei, Rend. Classe Sci. Fis. Mat. Nat. *8,* 494 (1950)
140) Caglioti, V., Liquori, A. M.: Atti Accad. Naz. Lincei, Rend. Classe Sci. Fis. Mat. Nat. *8,* 443 (1950)
141) Baker, L. C. W., Weakley, T. J. R.: J. Inorg. Nucl. Chem. *28,* 447 (1966)
142) Schaal, R., Souchay, P.: Anal. Chim. Acta *3,* 114 (1949)
143) Evans, H. T., Jr.: J. Am. Chem. Soc. *70,* 1291 (1948)
144) Evans, Jr., H. T.: J. Am. Chem. Soc. *90,* 3275 (1968)
145) Ripan, R., Calu, N.: Studia Univ., Babes-Bolyai, Ser. Chem. *10,* 135 (1965)
146) Jander, J., Jahr, K. F.: Kolloid-Beihefte *41,* 308 (1935)
147) Malaprade, L.: Ann. Chim. (Paris) *11,* 188 (1929)
148) Wiese, G.: Z. Naturforsch., Teil B *29,* 116 (1974)
149) Wiese, V., Boese, D.: Z. Naturforsch., Teil B *29,* 68 (1974)
150) Tsigdinos, G. A.: Ph. D. Thesis, Boston University, 1961, p. 29
151) Tsigdinos, G. A., Pope, M. T., Baker, C. W.: Abstracts of Papers Presented Before the Division of Inorganic Chemistry, Am. Chem. Soc., National Meeting Boston, 1959
152) Brown, D. H.: J. Inorg. Nucl. Chem. *14,* 129 (1960)
153) Chaveau, F., Souchay, P., Schaal, R.: Bull. Soc. Chim. Fr. *1959,* 1190
154) Scott, W. W.: Standard methods of chemical analysis. Furman, N. H. (ed.). New York: D. Van Nostrand Company 1938, Vol. I, p. 894
155) Tsigdinos, G. A.: Electrochemical properties of heteropoly molybdates. Climax Molybdenum Company Bulletin Cdb-15, May 1971
156) Hahn, H., Hahn, F.: Naturwissenschaften *49,* 539 (1962)
157) Hahn, H., Schmidt, G.: Naturwissenschaften *49,* 512 (1962)
158) Papaconstantinou, E., Pope, M. T.: Inorg. Chem. *6,* 1152 (1967)
159) Pope, M. T., Varga, G. M., Jr.: Inorg. Chem. *5,* 1249 (1966)
160) Pope, M. T., Papaconstantinou, E.: Inorg. Chem. *6,* 1147 (1967)
161) Kemula, W., Rosolowskii, S.: Rocz. Chem. *37,* 941 (1963)
162) Kemula, W., Rosolowskii, S.: Rocz. Chem. *38,* 905 (1964)
163) Meites, L.: Polarographic techniques. 2nd edit. New York: Interscience Publishers 1967, pp. 577–592
164) Headridge, J. B.: Electrochemical techniques for inorganic chemists. New York-London: Academic Press 1969, Chapter 5
165) Smith, D. E.: AC polarography and related techniques: theory and practice. Bard, A. J. (ed.). Electroanalytical chemistry, Vol. 1. New York: Marcel Dekker 1966, pp. 1–155

166) Duca, A., Budiu, T.: Rev. Roumaine Chim. *10,* 193 (1965)
167) Alimarin, I. P., Shakhova, F. F., Neotorkina, R. K.: Dokl. Akad. Nauk SSSR *106,* 61 (1956)
168) Bamann, E., Schiever, K., Freytag, A., Toussant, R.: Ann. Chim. (Paris) *605,* 65 (1957)
169) Souchay, P., Massart, R.: C. R. Acad. Sci., Paris, Ser. C, 253 (1961)
170) Massart, R., Souchay, P.: C. R. Acad. Sci., Paris, Ser. C, *256,* 4671 (1963)
171) Hahn, H., Becker, W.: Naturwissenschaften *49,* 513 (1962)
172) Hahn, H., Becker, W.: Naturwissenschaften *50,* 402 (1963)
173) Rabette, P., Olivier, D.: Rev. Chim. Minerale *7,* 181 (1970)
174) Rabette, P., Olivier, D.: C. R. Acad. Sci., Paris *277,* 559 (1973)
175) Rabette, P., Olivier, D., Pezerat, H.: Rev. Chim. Minerale *11,* 157 (1974)
176) Rabette, P., Olivier, D.: J. Less-Common Metals *36,* 299 (1974)
177) Rabette, P., Olivier, D.: Rev. Chim. Minerale *11,* 185 (1974)
178) Papaconstantinou, E., Pope, M. T.: Inorg. Chem. *9,* 667 (1970)
179) Rabette, P., Ropars, C., Grivet, J. P.: Compt. Rend. C*265,* 153 (1967)
180) Pope, M. T.: Inorg. Chem. *11,* 1973 (1974)
181) Evans, H. T., Showell, J. S.: J. Am. Chem. Soc. *91,* 6881 (1969)
182) Ama, T., Hidaka, J., Shimura, Y.: Bull. Chem. Soc. Jap. *43,* 2654 (1970)
183) Nishikawa, T., Sasaki, Y.: The University of Tokyo, private communication
184) Barkigia, K. M., Rajkovic, L. M., Pope, M. T., Quicksall, C. O.: J. Am. Chem. Soc. *97,* 4146 (1975)
185) Kokorin, A. I., Polotebnova, N. A.: Zh. Obshch. Khim. *26,* 3 (1956)
186) Williams, W. W., Conley, J. W.: Ind. Eng. Chem. *47,* 1507 (1955)
187) Matijevic, E., Kerker, M., Beyer, H., Theubert, F.: Inorg. Chem. 2, 581 (1963)
188) Polotebnova, N. A.: Zh. Neorg. Khim. *10,* 1498 (1965)
189) Polotebnova, N. A., Neimark, Y. L.: Zh. Neorgan. Khim. *11,* 1046 (1966)
190) Weakley, T. J. R., Malik, S. A.: J. Inorg. Nucl. Chem. *29,* 2935 (1967)
191) Tourne, C.: Compt. Rend. Ser. C, *266,* 702 (1968)
192) Tourne, C., Tourne, G.: Compt. Rend., Ser. C, *266,* 1363 (1968)
193) Sharpless, N. E., Munday, J. S.: Anal. Chem. *29,* 1619 (1957)
194) Nakamoto, K.: Infrared spectra of inorganic and coordination compounds. New York: John Wiley and Sons 1963, p. 107
195) Souchay, P., Dubois, P.: Ann. Chim. (Paris) *3,* 105 (1948)
196) Otake, M., Komiyama, Y., Otaki, T.: J. Phys. Chem. *77,* 2896 (1973)
197) Meyer, E., Jr., Huckfeldt, R.: J. Phys. Chem. *74,* 164 (1970)
198) Rabette, P., Ropars, C., Grivet, J. P.: Compt. Rend. Ser. C, *265,* 153 (1967)
199) Stonehart, P., Koren, J. G., Brinen, J. S.: Anal. Chim. Acta *40,* 65 (1968)
200) Rashkin, J. A., Pierron, E. D., Parker, D. L.: J. Phys. Chem. *71,* 1265 (1967)
201) Molybdenum Catalyst Bibliography (1950–1964): Climax Molybdenum Company, Compiled by P. O. Warner and H. F. Barry
202) Molybdenum Catalyst Bibliography (1965–1967), Supplement 1, Climax Molybdenum Company, Compiled by E. N. Losey and D. K. Means
203) Molybdenum Catalyst Bibliography (1968–1969), Supplement 2, Climax Molybdenum Company, Compiled by S. Rudolph
204) Molybdenum Catalyst Bibliography (1970–1973), Supplement 3, Climax Molybdenum Company, Compiled by G. A. Tsigdinos
205) McKinley, J. B.: Catalysis. Emmet, P. H. (ed.). New York: Reinhold 1957, Vol. V, pp. 405–526
206) Sheng, M. N., Zajecek, J. G.: Advan. Chem. Ser. No. 57, 418 (1968)
207) Shenderova, R. I., Bakhshizade, D. M. A., Smirnova, N. A.: Azerb. Khim. Zh. (*2*), 65 (1967)
208) Sebrulsky, R. T., Henke, A. M.: Ind. Eng. Chem., Process Des. Develop. *10,* 272 (1971)
209) British Patent 990, 639 (to Sosieta Edison, Settore Chimico), April 28, 1965
210) British Patent 1, 167, 732 (to ICI), October 22, 1969
211) Giordona, N., Caporali, G., Ferlazzo, N.: U.S. 3, 226, 421, December 28, 1965
212) Marisic, M. M.: J. Am. Chem. Soc. *62,* 2312 (1940)

213) Tanner, H. G. (to E. I. duPont de Nemours and Company): U.S. Patent 2, 173, 187, September 19, 1939
214) Saraf, V. A., Dole, K. K.: Indian J. Appl. Chem. *22,* 1 (1959)
215) Hargis, C. H., Young, H. S. (to Eastman Kodak Company): French Patent 1, 359, 141, April 24, 1964
216) Fetterly, L. C., Koetitz, K. F., Conklin, G. W.: French Patent 1, 342, 962, November 15, 1963
217) Klinkenberg, J. W. (to Shell Oil Company): U.S. Patent 2, 982, 799, May 2, 1961
218) Vol'pova, E. G., Oyloblina, L. I.: Tr. Groznensk, Neft. Nauchn.-Issled, Inst. (15), 265 (1963)
219) Klinkenberg, J. W., Waterman, H. I.: 5-yi [Pyatyi]-Mezhdunar Neft. Kongress, 1959; CA, *58,* 393d (1963)
220) Mazonski, T., Gasztych, D., Zielinski, W.: Przemysl Chem. *41,* 251 (1962)
221) Beres, T., Jakubowicz, L.: Przemysl Chem. *41,* 708 (1962)
222) Belgium Patent 619, 012 (to Stamicarbon, N. Y.): December 17, 1962; CA, *59,* 5072g (1963)
223) Callahan, J. L. (to Standard Oil Company of Ohio): Belgium Patent 625, 912, March 1963
224) Young, H. S., McDaniel, E. L. (to Eastman Kodak Company): U.S. Patent 3, 539, 309, June 15, 1964
225) McDaniel, E. L., Young, H. S. (to Eastman Kodak Company): French Patent 1, 409, 061, September 27, 1963
226) Kato, T., Aoshima, A., Kubota, Y., Matsumura, K. (to Asahi Chemical Industry Company, Ltd.): German Patent, 3, 345, 417, March 8, 1968
227) Cahoy, R. P., Coyne, D. M. (to Gulf Oil Corp.): U.S. Patent 3, 345, 417, October 3, 1967
228) Young, H. S. (to Eastman Kodak Company): U.S. Patent, 3, 379, 652, April 28, 1967
229) Kolchin, I. K., Gribov, A. M.: Khim. Prom. *44,* (2) 81 (1968)
230) Nakajima, H., Kominomi, N.: J. Catal. *8,* 197 (1967)
231) Scherhay, B., Hausweiler, A., Schwarzer, K., Stroth, R.: German Patent, 1, 235, 297, March 2, 1967
232) Borodkin, V. F., Usacheva, K. V.: Izv. Vysshykh Uchebn. Zavedenii, Khim. i Khim. Technol. *142,* (3), 1958; CA, *53,* 4295f (1959)
233) Moore, F. W.: Climax Molybdenum Company: unpublished results
234) Truax, T. R.: U.S. Forest Products Laboratory, R1118 (1935)
235) Truax, T. R., Harrison, C. A., Baechler, R. H.: Proc. Amer. Wood Preservers' Assoc. *1933,* 231
236) Amaro, A. J., Lipska, A. E.: Development and evaluation of practical self help fire retardants. Stanford Research Institute Report, August 1973
237) Gruntfest, I. J., Young, E. M., Jr.: Am. Chem. Soc. Div. Org. Coating Plastics Chem., Preprint 21, No. 1, 113 (1962)
238) Basher, D. M., Rhoades-Brown, J. E.: Brit. Corros. J., *8,* 50 (1973)
239) Robitaille, D. R.: Climax Molybdenum Company: unpublished results
240) Steinberger, L. (to American Products, Inc.): U.S. Patent 3, 009, 842, March 21, 1960
241) Roberts, G. L., Jr., Fessler, R. G. (to American Cyanamid Company): U.S. Patent 3, 365, 313, January 23, 1968
242) Roberts, G. L., Fessler, R. G. (to American Cyanamid Company): U.S. Patent 3, 346, 604, October 10, 1967
243) Lizlovs, E. A.: J. Electrochem. Soc. *114,* 1015 (1967)
244) Clarke, R. A. (to E. I. duPont de Nemours and Company): U.S. Patent 2, 173, 187, September 19, 1939
245) Willard, H. H., Diehl, H.: Advanced quantitative analysis. New York: D. Van Nostrand Company 1943, pp. 196–209
246) Wandelin, C., Mellon, M. G.: Anal. Chem. *25,* 1668 (1953)
247) Gullstrom, D., Mellon, M. G.: Anal. Chem. *25,* 1809 (1953)
248) Filippov, M. P., Gushchnina, L. F.: Zh. Anal. Khim. *19,* 480 (1964)
249) Strachoto, J., Kotasek, Z.: Chem. Listy *52,* 1093 (1958)
250) Deshpande, G. R., Karmarkar, S. S.: Hindustan Antibiotics Bull. *3,* 174 (1961)

251) Pelczar, T.: Dissertations Pharm. *12,* 107 (1960)
252) Matsui, K., Tazoe, K., Murakami, F.: Bitamin (Kyoto) *20,* 91 (1960)
253) Fujisaki, M., Arai, Y., Kon, T., Itoh, M., Mabashi, C.: Endocrinol. Japan *3,* 1 (1956)
254) Ming-Chue Huang, *et al.:* Yao Hsueh Pao *7,* 287 (1959)
255) Berisso, B.: Publ. Inst. Invest. Microquim., Univ. Nac. Litoral (Sec. Buenos Aires) *16* (16), 2 (1952)
256) Kuleshova, M. I.: Nekotonye Vopr Lekarstovoved *21,* 1959; CA*54,* 21637i (1960)
257) Yoshimura, K., Morita, M.: Bull. Natl. Hug. Lab, Tokyo *73,* 141 (1955)
258) Shroff, A. P., Arvin, P., Garner, J. K.: J. Chromatogr. Sci. *10,* 504 (1972)
259) Karina, L. M., Tsvestov, N. A.: Uch. Zap. Mosk. Inst. Tonkli Khim. Tekhnol. (*2*), 49 (1969)
260) Abdoh, Y.: Actes Congr. Sci. Intern. Tabac, 2^e Brussels *1958,* 499; CA, *55,* 21490g (1961)
261) Sauvagnac, R., Rosa, U.: Intern. Elettron Nucl., 7, Congr. Nucleare, 5th Rome *2,* 305 (196); CA, *56,* 8265 (1962)
262) Odell, N. R., Lyons, J. F. (to Texaco, Inc.): U.S. Patent 2, 297, 237, March 21, 1961
263) Marks, A. M.: Development of an inorganic phototropic system for flashblindness protection. Marks Polarized Corporation, Technical Report 67-2-CM (AD 639, 701) July 1966
264) Averbach, A. (to A. B. Dick Co.): U.S. 3, 623, 866, November 30, 1971
265) Morishima, T., Shimizu, R.: U.S. 3, 769, 043, October 30, 1973
266) Eriks, K., Yannoni, N. F., Agarwala, U. C., Simmons, V. E., Baker, L. C. W.: Acta Cryst. *13,* 1139 (1961)
267) Smith, P. M.: Ph. D. Thesis, Georgetown University, 1972
268) Hau, H. H.-K.: Ph. D. Thesis, Boston University, 1970
269) Evans, H. T.: Acta Cryst. B*30,* 2095 (1974)
270) Baker, L. C. W., Baker, V. S., Eriks, K., Pope, M. T., Shibata, M., Rollins, O., Fang, W., Koh, L. L.: J. Am. Chem. Soc. *88,* 2329 (1966)
271) Tsigdinos, G. A.: Heteropoly compounds. Methodicum chimicum. (English Edition), Niedenzu, K., Zimmer, H. (ed.). New York: Academic Press 1976, Vol. 8, Chapter 32
272) Matswuoto, K. Y., Kobayashi, A., Sasaki, Y.: Bull. Chem. Soc. Jap. *48,* 3146 (1975)
273) Nishikawa, T., Sasaki, Y.: Chem. Letters *1975,* 1185
274) Matsumoto, K. Y., Kato, M., Sasaki, Y.: Bull. Chem. Soc. Jap., *49,* 106 (1976)
275) Moore, F. W., Tsigdinos, G. A.: J. Less-Common Metals, *54,* 297 (1977)

Received December 5, 1977

Inorganic Sulfur Compounds of Molybdenum and Tungsten

Their Preparation, Structure, and Properties

George A. Tsigdinos

Climax Molybdenum Company of Michigan, Research Laboratory, A Subsidiary of AMAX Inc., Ann Arbor, Michigan 48105, USA

In Memory of my Wife Shirley

Table of Contents

Introduction

In recent years considerable interest has been shown in sulfur-containing compounds of molybdenum. Major interest has centered on the use of molybdenum-containing catalysts in the hydrodesulfurization of petroleum and in the use of molybdenum disulfide as a lubricant. Although sulfur can replace oxygen isomorphously in several compounds, such substitution when applied to molybdenum often leads to the formation of molybdenum-sulfur compounds entirely unrelated to those containing only oxygen. For example, considerable differences in structure, properties, and composition exist between the oxides and sulfides of molybdenum[1, 2]. Similar differences are also found between the oxyhalides and thiohalides. All too often, oxygen-containing molybdenum compounds have no corresponding sulfur analogs as in the case of the heteropoly[3] and isopoly compounds[4]. Only in the case of the simple molybdates has partial or total substitution of oxygen by sulfur been effected; the solution behavior of the thiomolybdate species is different from that of the corresponding oxygen analogs. The normal molybdates have been discussed elsewhere[5]. The thiomolybdates will be one of the topics discussed in the present chapter.

The older literature on molybdenum-sulfur chemistry describes several compounds[6] that have later been shown to be either incorrectly formulated or non-existent. In recent times, the chemistry of molybdenum-sulfur compounds has been elucidated by modern techniques, both in solution and in the solid state. It is therefore the purpose of the present chapter to review and systematize the best information available in this area. Although the primary emphasis has been placed on molybdenum-sulfur species, for completeness and correlation purposes, certain tungsten or selenium analogs have also been included.

Sulfides

While the existence of many molybdenum sulfides has been reported in the literature, only MoS_2, Mo_2S_3, and MoS_3 are well established. The existence of molybdenum pentasulfide, Mo_2S_5, is still uncertain[7]. It is doubtful that the molybdenum tetrasulfide, MoS_4, reported in the early literature[6] to have been prepared from aqueous solutions, exists. No evidence was obtained for the presence of molybdenum monosulfide in phase studies of the binary Mo–S system[8]. A new sulfide Mo_3S_4 has been prepared and characterized[222].

Molybdenum Disulfide

Occurrence

Molybdenum disulfide or molybdenite (MoS_2) is the principal natural source of molybdenum and is widely distributed throughout the world[2]. The largest world deposits of molybdenite are in Colorado. Of the thirteen minerals in which molybdenum has been found[2], wulfenite ($PbMoO_4$) and molybdenite are present in sufficient concentration to be considered as commercial ores. Molybdenum production outside of Colorado usually accompanies that of associated metals such as copper and tungsten.

Although the molybdenite obtained from commercial ores is hexagonal, a rhombohedral modification of MoS_2 has also been found in nature[9]. Polytypism in molybdenite and the distribution of naturally occurring polytypes of this mineral have been discussed in the literature[9–11].

Preparation (Hexagonal Form)

Hexagonal molybdenum disulfide can be prepared directly from the elements by heating, or by thermal decomposition of ammonium tetrathiomolybdate or molybdenum trisulfide. Reports that hexagonal MoS_2 can be prepared by heating molybdenum trioxide in hydrogen sulfide[12] are probably incorrect. Recent evidence indicates that bulk molybdenum trioxide reacts with hydrogen sulfide at 400–500 °C[13], or with H_2S/H_2 mixtures at 300–500 °C[14], to yield mixtures of MoO_2 and MoS_2. Further sulfiding results in the slow conversion of MoO_2 to MoS_2[14]. All of these methods of preparation result in the formation of the hexagonal form only; no evidence for the rhombohedral form was obtained[15]. Samples of molybdenum disulfide prepared by these methods require prolonged annealing at 1100 °C to obtain well-crystallized hexagonal MoS_2 with sharp X-ray powder lines[15]. The annealing treatment removes stacking faults found in synthetic molybdenum disulfide[15]. The preparation of molybdenum disulfide directly from the elements is carried out at ≈ 1100 °C[16]. Methods of growing single crystals of MoS_2 by direct vapor transport have been described[17].

The preparation of crystalline molybdenum disulfide by the thermal decomposition of molybdenum trisulfide was studied by X-ray diffraction, and a model for the crystallization process was proposed[15]. Thus, MoS_3 loses sulfur when heated above 250 C, but the products remain amorphous up to ≈ 350 °C. Crystallinity sets in at ≈ 400 °C when the first vague lines are observed in the X-ray diagrams. At ≈ 500 °C, new broad lines occur which, upon further heating, become sharper. Only prolonged heating at 1100 °C results in haxagonal MoS_2 having crystallinity comparable to that of natural molybdenite[17]. Similar observations were also obtained in the study of the thermal decomposition of ammonium tetrathiomolybdate, $(NH_4)_2MoS_4$[18], to yield MoS_2 above 400 °C. However, the reported thermal decomposition of MoS_3 at 500–600 °C to yield first rhombohedral MoS_2, which upon further heating is converted to the hexagonal form[19], has been questioned[15]. The original interpretation was based on misassignment of X-ray powder lines to the ordered rhombohedral form that were actually due to disordered stacking in the hexagonal layered structure[15]. However, the hydrogen reduction of MoS_3 has been shown to yield both forms of MoS_2, the rhombohedral modification formed only at a reduction temperature of about 800 °C[20, 21].

A model has been proposed which describes the observations which occur during the thermal decomposition of MoS_3 and $(NH_4)_2MoS_4$[15], i.e., the appearance of lines in the X-ray powder diagrams resulting from the crystallization of MoS_2. During thermal decomposition of MoS_3 or $(NH_4)_2MoS_4$, gaseous sulfur, hydrogen sulfide, or ammonia is liberated, and the products resulting at 400 °C are very porous. Indeed, when MoS_3 is heated above 250 °C, it loses sulfur, but the products remaining are amorphous up to about 350 °C. Crystallization begins gradually above this tempera-

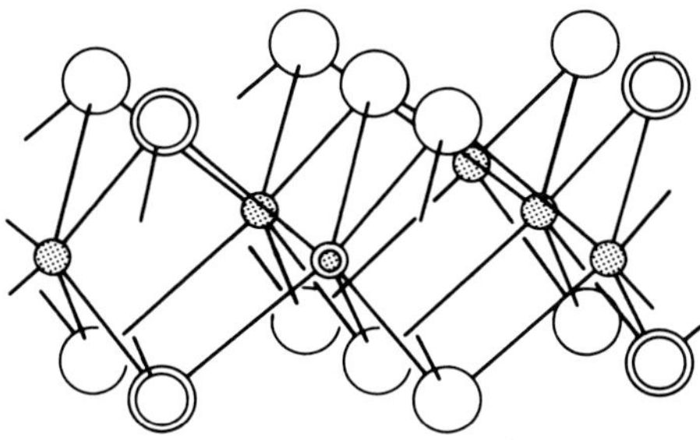

Fig. 1. Fragment of a layer of MoS_2
The metal atoms are indicated by small shaded circles, the sulfur atoms by open circles. The atoms drawn with double circles lie in one (110) plane

Fig. 2. Completely random orientation of the one-layer-crystallites; powder pattern shows (hkO) bands only

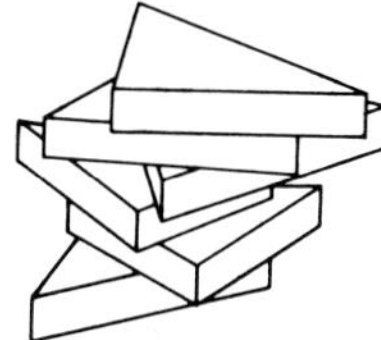

Fig. 3. Stacking of the layers in random orientations about the c-axis; (hkO) and (001) bands

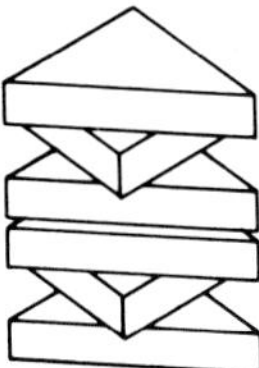

Fig. 4. Hexagonal MoS_2 with stacking faults; sharp lines for $h-k = 3n$, lines broadened if $h-k \neq 3n$

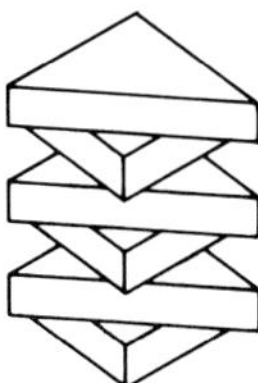

Fig. 5. Hexagonal MoS_2; Adjacent layers in anti-parallel orientation; sharp powder lines

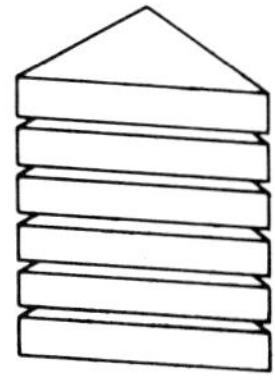

Fig. 6. Rhombohedral MoS_2; all layers in parallel orientation; sharp powder lines

Figs. 2–6. A model for the crystallization of $Mo_{1-x}S_2$
[Triangles depict fragments of $Mo_{1-x}S_2$ layers]

ture. The composition of the product at 350 °C is $Mo_{0.83}S_2$. These crystallites, that may be regarded as fragments of MoS_2 layers (see Fig. 1), are randomly oriented as shown in Fig. 2. Further heating results in their growth by the loss of sulfur and further orientation. Complete crystallinity is finally obtained by heating at 1100 °C. The crystallization process described is shown in Figs. 2 through 6.

Preparation (Rhombohedral Form)

Rhombohedral MoS_2 is prepared by the reaction of molybdenum trioxide and sulfur in molten potassium carbonate at 900 °C[22]. Well-crystallized material was also prepared in nearly quantitative yield employing sodium carbonate instead of the potassium salt[15]. Prolonged heating of rhombohedral MoS_2 at 1000 °C under vacuum leads to conversion into the hexagonal form[15].

The ultrahigh-pressure, high-temperature synthesis of rhombohedral dichalcogenides of molybdenum and tungsten has been described[23]. When a 1:2 elemental molybdenum metal-sulfur mixture was hot pressed at 27 kbars at 900 °C, or 47 kbars at 800 °C, only hexagonal molybdenite formed after a five-minute treatment. However, at 47 kbars and 1050 °C, 74 kbars and 1100 °C, or up to 70 kbars and 2000 °C, quantitative yields of the rhombohedral form were obtained. Normal hexagonal molybdenite is completely transformed to the rhombohedral form at 40–75 kbars at 1900–2000 °C in 1 to 5 minutes. Apparently, the synthesis of MoS_2 from its elements is mainly temperature dependent. The starting mixture stoichiometry also determines the final product. Molybdenum to sulfur mixtures with 1:1 atomic ratio produce only Mo_2S_3 over the pressure and temperature ranges of 10–75 kbars and 800–2200 °C.

Rhombohedral molybdenum disulfide was synthesized in an autoclave under hydrothermal conditions[24]. Sodium molybdate was converted to MoS_4^{-2} with hydrogen sulfide and then heated. Amorphous MoS_2 was formed at 20–300 °C and a solution pH $\leq$ 6–7; colloid-like MoS_2 formed at 200–300 °C; rhombohedral MoS_2 formed at 250–900 °C in a few hours; and the hexagonal modification formed after 22 days at 600 °C, or two hours at 1300 °C. In addition, the hydrothermal crystallization of molybdenite was studied using thiomolybdate solutions with a general formula $M_2(MoS_{4-x}O_x)$ where M is an alkali metal. The synthesis was carried out in an autoclave at 100–600 C and $<$ 800 kg/cm^2 pressure. The thermal dissociation of the solution resulted in the formation of two modifications of MoS_2 hexagonal (2H) and rhombohedral (3R), as ascertained by electron diffraction studies[25].

Other Methods for Preparing MoS_2

Several other methods for preparing molybdenum disulfide have been reported in the literature, but caution should be exercised in assigning a particular crystal modification to the materials obtained on the basis of powder diagrams[15]. Such materials often show stacking faults; thereby, the powder patterns show differences from those of natural molybdenite[15].

Hexagonal molybdenite was formed at 1300 C in a silicate melt containing molybdenum metal, sulfur, and sodium chloride[24]. Pure molybdenum disulfide of stoichiometric composition was also prepared by the reaction of sulfur with molybdenum trioxide or calcium molybdate in fused sodium chloride or molten sodium or potassium carbonate, the material thus obtained having lubricating properties reportedly equal to those of the natural product[26, 27]. Molybdenum disulfide has also been prepared from fused salt electrolysis[28, 29], carried out at 800 °C and involved mixtures of molybdenum trioxide, sodium borate, sodium fluoride, and sodium sulfate[28]. The reaction of anhydrous sodium molybdate and sulfur at 500–700 °C has been reported to yield molybdenum disulfide[30–32]. Other methods of preparation include the direct reaction of sulfur with calcium molybdate, molybdenum pentachloride, molybdenum trioxide, or molybdenum metal; the synthetic MoS_2 thus produced reportedly having lubricating properties comparable to natural molybdenite[29]. The reaction of hydrogen sulfide with molybdenum metal at 640–700 °C and at low pressure yields MoS_2 with a lamellar structure[33]. The electrosynthesis of stable molybdenum sulfide sols has also been described[34].

Crystal Structures of MoS_2

Molybdenum disulfide exists in two modifications: (1) the well-known hexagonal form (a = 3.16, c = 12.29 Å) and (2) the rhombohedral form (a = 3.17, c = 18.38 Å). The crystal structure of molybdenum disulfide was first investigated in 1923 by Dickinson and Pauling on natural molybdenite[35]. The compound is hexagonal and characterized by MoS_2 layers in which the Mo atoms have trigonal prismatic coordination of six sulfur atoms. There are two molecules per unit cell. A fragment of the MoS_2 layer is shown in Fig. 1. The crystal structure of MoS_2 is shown in Fig. 7. The distance between molybdenum atoms and the nearest sulfur atoms is 2.41 ± 0.06 Å. The edge of the MoS_6 prism is the thickness of the MoS_2 layer, 3.15 ± 0.02 Å. The distance between the two adjacent sulfur layers is 3.49 Å, which is greater than the thickness of the layers themselves.

In 1957 Bell and Herfert[22] synthesized a rhombohedral modification of MoS_2 and showed it to have a similar layer structure of the Mo atoms. By qualitative study of (001) reflections, an octahedral coordination was assigned. However, it was shown later that the coordination about Mo in rhombohedral MoS_2 is also trigonal prismatic[36]. Since then, rhombohedral MoS_2 has been found to occur in nature in the Mackenzie District of Canada[37], in Switzerland[38], in Portugal[39], and in the USSR[40, 41].

Rhombohedral and hexagonal structures are different only in the method of stacking of the MoS_2 layers. A detailed investigation of the crystal structure of rhombohedral MoS_2[9, 42] and a systematic deduction of possible polytypes of molybdenite have been carried out[9]. Possible polytypes with simple structures derived theoretically gave one rhombohedral (3R), two hexagonal ($2H_1$, $2H_2$), and one trigonal (2T) crystal forms. Here, the R, H and T stand for the basic crystal structures, the coefficient represents the number of molecules per unit cell, and the subscript is used to distinguish between different arrangements within the same class

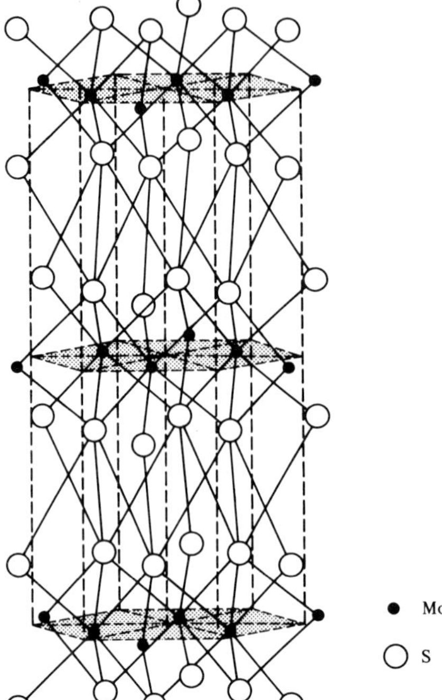

Fig. 7. Crystal structure of MoS_2

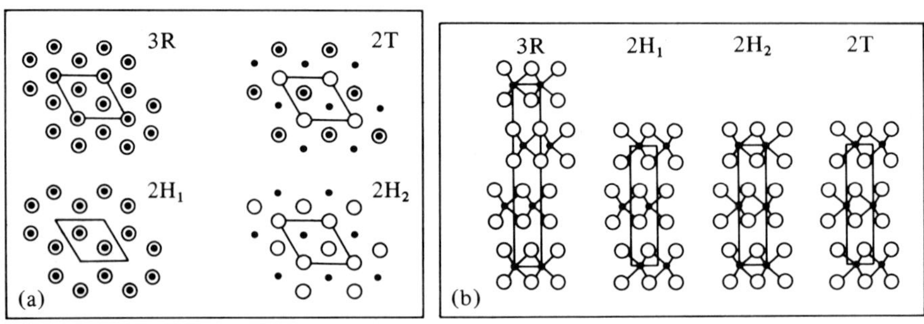

Fig. 8. Projections of MoS_2 crystal polytypes (after Takeuchi and Nowacki[9])
(a) Structures of simple polytypes along the c-axis; (b) structures of simple polytypes projected along the a_2 axis. Note: Figs. 8(a) and 8(b) can be thought of as plan and elevation views respectively of the various MoS_2 simple polytype structures if the MoS_2 platlet were resting with its base plane on a horizontal plane. The solid circles represent the molybdenum atoms, and the open circles represent the sulfur atoms. In Fig. 8(a), a solid circle inside an open circle represents the Mo and S atoms situated one above the other

of crystal structures (see Fig. 8). In addition, the 112 theoretically possible polytypes of molybdenite with less than seven layers have been derived[10]. Of the 108 specimens of naturally-occurring molybdenite studied, representing 83 localities throughout the word, 80 percent of these were the $2H_1$ polytype, three were the 3R polytype, and the remainder were mixtures of $2H_1$ and 3R in varying proportions[11]. No new modifications of naturally-occurring molybdenite were found[11]. So far, only the 3R and $2H_1$ polytypes have been definitely identified in natural or in synthetic MoS_2. Projections of the atoms for the four polytypes 3R, $2H_1$, $2H_2$ and 3T are shown in Fig. 8.

The trigonal prismatic bond orbital configuration in MoS_2 has been discussed[43]. X-ray absorption and emission L spectra of this compound have also been obtained[44, 45]. The results are consistent with a d^4sp hybridization in MoS_2.

Properties

The properties of molybdenum disulfide have been given elsewhere[12]. However, an updated version and recent references to the literature will be given here.

Tab. 1. Physical properties of MoS_2

Molecular Weight	160.08
Color	Blue-Gray to Black
Melting Point	Decomposition[1)]
Specific Gravity	4.8 to 5.0
Crystal Structure	Hexagonal (usual form) or Rhombohedral. Each Mo atom is surrounded by a trigonal prism of S atoms
Hardness	1, Mohs Scale; 12–60 Knoop
Magnetism	Diamagnetic

1) In vacuum to Mo_2S_3; in air to MoO_3.

Physical Properties. The pertinent physical properties of molybdenum disulfide are given in Table 1. Although a melting point of 1185 °C for MoS_2 has been reported in the older literature[12], it is most likely incorrect[46]. No melting was observed when molybdenum disulfide was heated under high vacuum at 1800 ± 20 °C for 10 minutes[46], although it is doubtful that molybdenum disulfide remains intact at those temperature[47]. Also, no melting of this material was observed at 1600 °C under 1 atmosphere of hydrogen, although decomposition into molybdenum metal and sulfur was observed[48].

Electrical, Magnetic, Optical, and Thermal Expansion Properties. The electrical and magnetic properties of MoS_2 have been summarized elsewhere[12]. Molybdenum disulfide is diamagnetic and a semiconductor. Recent work on the electrical properties of MoS_2 is given in Refs. [49] through [51]. Electron spin resonance measurements on molybdenum disulfide suggest that the signal obtained is due to mobile charge carriers rather than to free radical centers[52].

The optical properties and photoconductivity of thin crystals of molybdenum disulfide have been investigated[53–58]. The thermal expansion properties of MoS_2 have also been reported[59]. The coefficient of thermal expansion is 10.7×10^{-6}/deg K.

Thermal Conductivity. The thermal conductivity of molybdenum disulfide in the temperature range of –315° F (between liquid-oxygen and liquid-nitrogen boiling temperatures) to 1090 °F (above the service temperature of MoS_2 in air) has been determined[60]. The data were determined with a guarded double-cylinder apparatus which provides an accuracy of ±10% at –315 °F and ±5% at 90 °F and the tested specimens were maintained at a pressure of 10^{-1} torr throughout the cooling and heating cycle. The thermal conductivity values for all the MoS_2 specimens evaluated between –315 °F and 1090 °F are within the range of 0.020 and 0.085 Btu/hr/in./°F. A graphical presentation of the data is given in Fig. 9.

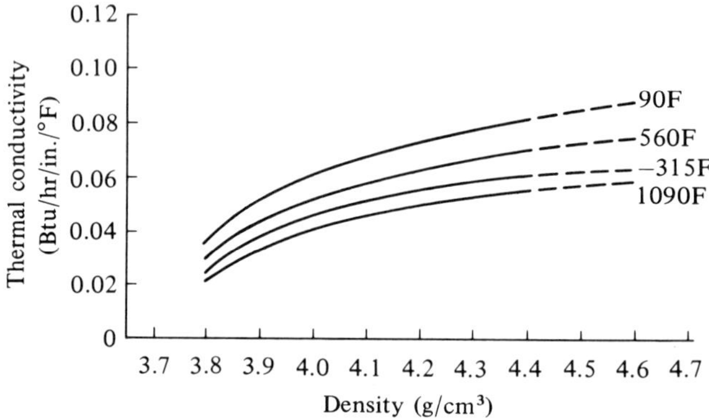

Fig. 9. Thermal conductivity of compacted MoS_2 as function of density for various temperatures

Thermodynamic Properties. Low temperature heat capacity and entropy data on MoS_2 have been developed[61]. The heat capacity exhibits approximately T^2 dependence between 20 and 60 K. For MoS_2 (c), an entropy of $S^0 298 = 14.96 \pm 0.02$ cal deg^{-1} $mole^{-1}$ has been reported[62]. The values for the heat of formation, $\Delta H^0 = -56.1$ kcals/mole, and for the free energy of formation, $\Delta G_f^0 = -54.1$ kcals/mole, have been calculated[63].

Thermal Behavior. The thermal behavior of molybdenum disulfide and its dissociation products under heating have been extensively investigated. Earlier studies on the equilibrium phases during the reduction of MoS_2 with hydrogen up to 1000 °C indicated that no lower sulfide than MoS_2 existed[64]; however, more recent work established that Mo_2S_3 and not MoS_2 is in equilibrium with molybdenum metal in the vicinity of 1100 °C[65]. High purity molybdenum metal can be produced by the direct dissociation of molybdenum disulfide in vacuum at 1600–1700 °C[66]. In practice, the dissociation temperatures are above 1370 °C, and the sulfur is continuously withdrawn to permit the reaction to proceed to completion. Such dissociation occurs in two discrete steps *via* the formation of Mo_2S_3 and S_2 [66]. Further studies of the equilibrium Mo_2S_3-sulfur at 1100 °C indicate that S_2 is the predominant molecular species at 1100 °C and high pressures; the Mo_sS_3 phase obtained was found to be stoichiometric [67]. The equilibria that exist between the sulfides of molybdenum and mixtures of hydrogen and hydrogen sulfide have also been studied[16], but they will be discussed later.

The dissociation of MoS_2 at normal pressures and in reducing atmosphere at 1500 °C proceeds to Mo_2S_3 [68,69]. In vacuum the dissociation proceeds at lower temperatures[47,68,69]. Mo_2S_3 begins to dissociate above 1500 °C, but at a pressure of 1 mm Hg, it dissociates at 1250–1300 °C[47,68,69]. At higher vacuum ($\sim 10^{-2}$ to 10^{-4} mm Hg), dissociation of Mo_2S_3 to the metal takes place at 1100–1200 °C[67,68] When MoS_2 was heated in an evacuated (1×10^{-5} mm Hg) quartz tube at 1280–1300 °C, Mo_2S_3 was obtained free of molybdenum metal[70].

Chemical Properties. Molybdenum disulfide is generally unreactive chemically. Molybdenite is converted, on a commercial scale, to molybdenum trioxide by roasting in air at 500–600 °C. The oxidation proceeds *via* highly exothermic reactions[71]. The kinetics of the oxidation of molybdenite have been studied in the temperature range 400–650 °C [71–73]. The oxidation proceeds *via* intermediate MoO_2; the formation of MoO_3 does not proceed to completion until all sulfur has been removed from the solid[71]. The oxidation of molybdenum disulfide lubricant at several temperatures has been reported[74,76]. The oxidation of MoS_2 to MoO_3 was studied by thermogravimetric analysis at 435, 466, and 516 °C; the data indicated the formation of 10, 50, and 90% MoO_3, respectively. The kinetics of the oxidation of synthetic and natural MoS_2 have been compared[77]. Higher temperatures were required for the oxidation of natural MoS_2, the oxidation process in this case being highly dependent on particle size[77].

Molybdenum disulfide dissolves in strong oxidizing agents such as *aqua regia*, hot concentrated sulfuric and nitric acids to give soluble hexavalent molybdenum species. With nitric acid, the degree of oxidation of MoS_2 to $MoO_3 \cdot xH_2O$ increases with the concentration of HNO_3 up to a certain limit typical of each temperature. The effect on oxidation was not significant for concentrations of HNO_3 above 30%[78]. The oxidation of molybdenite by potassium permanganate in sodium carbonate has been examined, the process being limited by diffusion[79]. Molybdenite is also oxidized to Na_2MoO_4 in sodium hydroxide solution by cupric oxide at 250–290 °C[80,81]. The dissolution of MoS_2 in alkaline solution was examined over the temperature range of 100–175 °C and pressure (of oxygen gas) range 0 to 700 psig. The rate of leaching by potassium hydroxide was a linear function of oxygen over-pressure and concentration of potassium hydroxide[82].

The oxidation of MoS_2 with chlorine monofluoride proceeds smoothly at 25 °C to yield MoF_6 [83]. Excess chlorine gas reacts with natural molybdenite at 550–600 °C to yield as main products the volatile MoO_2Cl_2 and $MoOCl_4$. This process has been proposed for the recovery of molybdenum from marginal ores[84]. Carbon dioxide begins to react with natural molybdenite at 200–250 °C; the maximum reaction rate is attained at 900–1000 °C to yield MoO_2, SO_2, CO, S, and some MoO_3 and COS[85]. A low carbon, low oxygen molybdenum metal is produced by the reduction of MoS_2 with tin[86]. No nitrides were formed under the conditions studied when MoS_2 was heated in nitrogen[87]. Reduction of MoS_2 with hydrogen gas proceeds to the metal *via* the formation of Mo_2S_3 [16]. During oxidative roasting of molybdenum concentrates, rhenium oxides react with MoS_2 to form ReO_2 when rhenium is present in the ore[88].

Molybdenum disulfide is not soluble in ordinary solvents. However, sodium hydrochlorite solutions oxidize MoS_2 slowly[89]. The solubility of MoS_2 was determined in sodium sulfide solutions at 60 °C and found to be very slight (0.89×10^{-5} mole/l)[90]. The solubility of MoS_2 was determined electrochemically using molybdenite electrodes; weak dc currents increased the solubility; increases in pH generally lowered the solubility[91]. Molybdenum disulfide dissolves in potassium cyanide by complex formation[12].

Intercalation Compounds of MoS_2. Like graphite, the lamellar structure of molybdenum disulfide permits it to form interlamellar compounds. For example, powdered MoS_2 adsorbs ammonia from –63.5 to –78 °C, the maximum amount of ammonia taken up corresponding to 50 mole percent[92]. The adsorption is reversible and without hysteresis, although attainment of equilibrium on desorption is slow. No change in X-ray parameters or expansion of the lattice was found during the sorption of ammonia. Thus there is appreciable room within the MoS_2 lattice to incorporate the ammonia molecules. This is corroborated by the fact that MoS_2 specimens, when subjected to very high pressures (up to 120 katms) at room temperature, undergo a linear compression to 60% of the original value. Re-expansion upon reducing the pressure is almost without hysteresis. The presence of ammonia between the planes is postulated to involve the presence of NH_4^+ and NH_2^- ions[92].

Other intercalation compounds of molybdenum disulfide and related species have been reported. Like graphite, MoS_2, $MoSe_2$ and WS_2 react with metal dissolved in liquid ammonia to form intercalation compounds, the magnetic behavior of which indicates that while these compounds have metallic character, there is a transition toward the formation of ionic sulfur or selenium bonds[93–95]. The intercalation compounds reported are given in Table 2. The compounds are prepared by the reaction of molybdenum disulfide or the other chalcogenides at –40 to –50 °C with liquid ammonia in which the specified metal has been dissolved[93]. The products

Tab. 2. Intercalation compounds of MoS_2, WS_2, $MoSe_2$ and WSe_2

		Parent Lattice	
MoS_2	$MoSe_2$ or $MoTe_2$	WS_2	WSe_2
$Cs_{0.5}MoS_2$	$K_{0.5}MoSe_2$	$Cs_{0.5}WS_2$	$K_{0.4}WSe_2$
$RbMoS_2$	$Eu_{0.7}MoSe_2$	$Rb_{0.5}WS_2$	$Eu_{0.5}WSe_2$
$K_{0.6}MoS_2$	$Eu_{0.4}MoTe_2$	$K_{0.5}WS_2$	$Eu_{0.4}WSe_2$
$Na_{0.6}MoS_2$	$Eu_{0.5}(NH_3)_{0.5}MoS_2$	$Na_{0.5}WS_2$	
$Na_{0.8}(NH_3)_{0.2}MoS_2$		$Li_{0.5}(NH_3)_{0.6}WS_2$	
$Li_{0.8}(NH_3)_{0.8}MoS_2$		$Na_{0.6}(NH_3)_{0.1}WS_2$	
$Li_{1.1}(NH_3)_{0.6}MoS_2$		$Ca_{0.6}(NH_3)_{0.6}WS_2$	

are washed with ammonia and dried in vacuum at room temperature. Although they are diamagnetic like the parent compounds, incorporation of the alkali metals into the lattice decreases the diamagnetism[94]. It is postulated that polar species like $M^+(MoS_2)^-$ exist within the intercalated compounds. The metallic behavior of the compounds is consistent with their pyrophoric properties or reaction with water to yield the parent sulfide, metal hydroxide, and hydrogen gas. No hydrogen sulfide is formed during hydrolysis[94]. Analysis indicates the structure consists of alternating layers of metal and matrix material[94, 95]. It has been shown that Ni^{+2} or Co^{+2} ions can be intercalated between the MoS_2 layers in octahedral holes situated adjacent to the exposed molybdenum atoms. The materials obtained are catalysts in hydrogenation of benzene[96].

Molybdenum disulfide intercalation compounds have been considered for application in electrical energy storage superbatteries[223–225].

Lubricating Properties. Molybdenum disulfide owes its lubricating properties primarily to its lamellar molecular structure (see Fig. 7); coefficients of friction of this material as low as 0.017 have been measured[12]. One study examined the variation of the coefficient of friction with crystal orientation; it was found that the minimum friction occurs when the basal plane is parallel to the sliding surfaces and maximum when the basal plane is perpendicular to the sliding surfaces[97]. Upgraded natural molybdenite is used commercially as a solid lubricant. It can be applied as a burnished film, as a bonded coating, or mixed with grease or suspended in oil. Detailed discussions of the lubricating properties of molybdenum disulfide are to be found in Refs.[98–102].

Catalytic Properties. Molybdenum disulfide is used as a catalyst in a variety of hydrogenation-dehydrogenation reactions involving complex hydrocarbon mixtures such as petroleum and coal tars. Detailed lists of the reactions where MoS_2 acts as a catalyst have been compiled[103–107]. For maximum surface area, the MoS_2 is usually prepared on the carrier *via* thermal decomposition of ammonium tetrathiomolybdate or reduction of ammonium molybdate with hydrogen sulfide. High surface area MoS_2 was prepared from MoS_3 at 450 °C by fast reduction with hydrogen or by thermal decomposition in helium. Surface areas as high as 158 m^2/g were thus obtained[108].

Molybdenum Sesquisulfide

Molybdenum sesquisulfide, Mo_2S_3, is the lowest sulfide in the Mo-S system. Contrary to previous claims[64], it has now been established[65] that Mo_2S_3, and not MoS_2, is in equilibrium with molybdenum metal and sulfur vapor in the vicinity of 1100 °C. The presence of Mo_2S_3 has also been established in equilibrium studies in the molybdenum-sulfur-hydrogen system between 850 and 1200 °C[16].

Preparation

Molybdenum sesquisulfide has been obtained by heating MoS_2 in an arc furnace at atmospheric pressure in the absence of air[2,109]. The product was separated from the reaction mass by treating with *aqua regia*. This material has also been prepared by heating molybdenum metal and sulfur (2 : 3 atomic ratio) in evacuated quartz ampoules at 1300–1400 °C[110–112].

Properties

The sesquisulfide is inert towards cold, dilute *aqua regia*, and it is not attacked by concentrated hydrochloric or sulfuric acids; Mo_2S_3 is readily converted to MoO_3 by warm concentrated nitric acid[2]. Mo_2S_3 reacts with sulfur vapors at high temperatures to form MoS_2[76]. The physical properties of Mo_2S_3 are given in Table 3.

Tab. 3. Physical properties of Mo_2S_3

Formula Weight	288.08
Color	Steel-Gray
Crystalline Form	Needles; Monoclinic, two Mo_2S_3 units per unit cell; Calculated Density = 5.806 g/cc[1])
Structure	Distorted octahedral configuration of S atoms about Mo. Below –80 °C, Mo_2S_3 structure distorts to triclinic symmetry[2])
Microhardness	350 to 510 kg/mm^2 [3])

[1]) Refs. [121] and [122].
[2]) Ref. [122].
[3]) Ref. [123].

The standard free energy of formation of Mo_2S_3 has been determined[113] and found to follow the equation:

$$\Delta G_f^0(1/3\ Mo_2S_3) = -41{,}700\ (\pm 1000) + 17.3\ (\pm 0.6)T \pm 300\ \text{cal}$$

between 1092 and 1337 °C. Similar results have been obtained in another study[16]. A computer estimation of the standard heat of formation of Mo_2S_3, ΔH_f^0, has yielded a value equal to –53 kcal/mole ±15[114]. The refractory nature of Mo_2S_3 has been used for impregnating porous ceramics with this material[115].

Structure

The structure of molybdenum sesquisulfide has been determined[111] on a sample of composition $Mo_{2.06}S_3$ prepared in vacuum at 1300 °C from molybdenum and sulfur. An electron density projection[110] of Mo_2S_3 is shown in Fig. 10. The Mo_2S_3 thus examined is monoclinic, with a = 6.092 Å; b = 3.208 Å; c = 8.6335 Å; β = 102.43°. Below +37 C, the structure is a superstructure of the one described, all axes being doubled. At –80 °C, the lattice of $Mo_{2.06}S_3$ undergoes a distortion to triclinic symmetry. As shown in Fig. 10, the molybdenum atoms are in octahedral holes of sulfur atoms but are displaced from the octahedron centers in such a way that

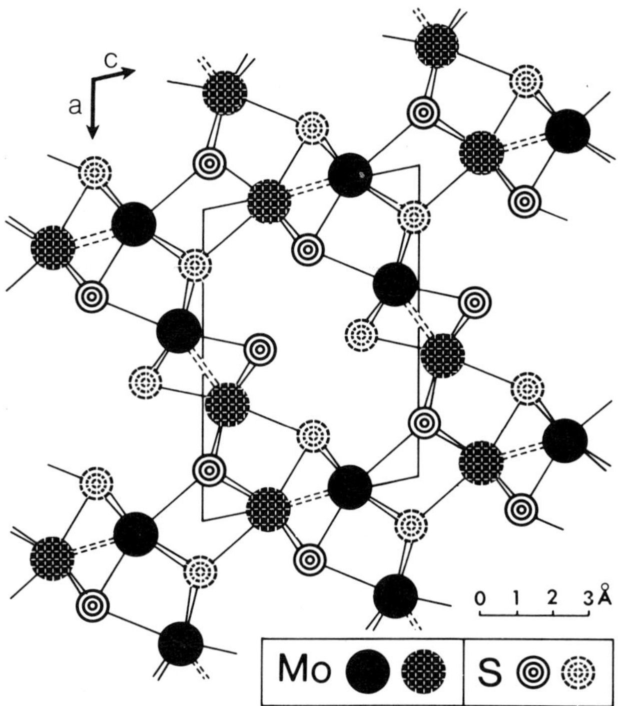

Fig. 10. Electron density map of Mo_2S_3
[Mo–S bonds are indicated by full lines, the Mo–Mo chains by broken lines. The outline of the unit cell is also indicated.]

infinite zig-zag Mo-Mo chains are formed. The Mo-Mo distances in these chains are 2.86 Å, which are only slightly longer than those present in molybdenum metal (2.725 Å). The lack of chemical reactivity of Mo_2S_3 is consistent with its structure. An orthorhombic modification of Mo_2S_3 that has been reported[112)] has been shown to be incorrectly assigned[110)]. Only the monoclinic form exists.

Up to one half of the molybdenum atoms in Mo_2S_3 can be replaced by niobium[111)]. The solid solutions of $(Mo_{1-x}, Nb_x)_{2.06}S_3$, the high temperature crystal structure of $Mo_{2.06}S_3$ is retained down to low temperatures if $0.025 < x \leqslant 0.50$[111)].

Thermal Behavior

Molybdenum sesquisulfide decomposes to the metal and sulfur when heated above 1600 °C, the rate of decomposition depending on the rate of removal of sulfur and the pressure of the system[66)]. Thermogravimetric analysis of Mo_2S_3 in a nitrogen atmosphere showed no weight change up to 800 °C and a very slight (0.5%) weight gain in going to 1200 °C[7)]. However, when the TGA was carried out in air (see Fig. 11), no weight change occurred up to 350 °C, where a very sharp weight loss occurred up to 430 °C; thereafter, the sample attained almost its original weight. This change is associated with the conversion of Mo_2S_3 to 2 MoO_3, a process ulti-

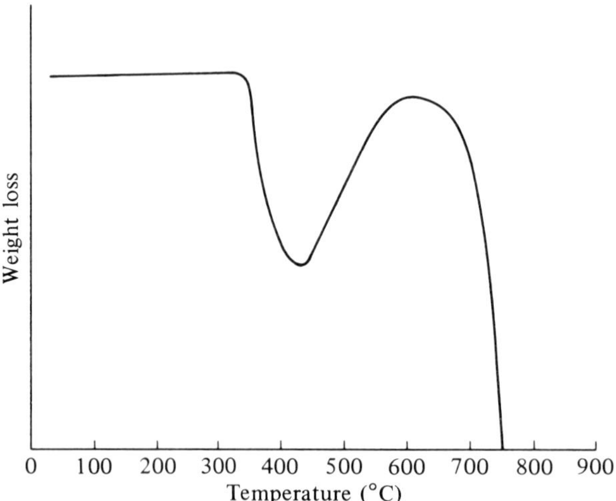

Fig. 11. Thermal Behaviour of Mo_2S_3 in Air

mately associated with no weight change. Further weight loss beyond 650 C corresponds to the volatilization of molybdenum trioxide[7].

Molybdenum Trisulfide

Since the precipitation of molybdenum as the trisulfide is a classical analytical separation technique, a considerable amount of chemistry has been reported on MoS_3. Although this compound has not been used for a gravimetric technique because of the variable composition of the precipitate, it has been reported that a compound having the exact composition $MoS_3 \cdot 2\,H_2O$ can be prepared[116,117]. It is also reported that the compound loses its water of hydration above 200 C but does not oxidize below 390 °C[19,117]. More definitive work[19] however indicates that regardless of the method of preparation from solution, the composition of molybdenum trisulfide was $MoS_{3+x} \cdot yH_2O$, where $0 < x < 1$ and $y > 0$. This excess sulfur cannot be washed out with carbon disulfide. Other work also indicates that molybdenum trisulfide precipitated from solution has variable composition such as $3\,MoS_3 \cdot H_2S \cdot H_2O$[118], or $MoS_3 \cdot H_2O$[119].

Preparation

Molybdenum trisulfide can be prepared by the thermal decomposition of $(NH_4)_2MoS_4$ in an inert atmosphere[7,15,19], or by heating the ammonium piperidinium and piperizinium tetrathiomolybdates at 200 °C in vacuum[15]. The "crystalline MoS_3" reported in the literature[120] was based on microscopic evidence only. So far, no real X-ray evidence has been found for the existence of a crystalline form of molybdenum trisulfide[15,21]. It has been suggested that at present there is no conclusive evidence which shows that MoS_3 is a definite chemical compound rather than an intimate mixture of subcrystalline MoS_2 and amorphous sulfur[21)].

Thermal Behavior

The thermal decomposition of molybdenum trisulfide to the disulfide is irreversible[121)]. Thermal studies on MoS_3 under inert atmosphere conditions show that it begins to lose sulfur at ≈250 °C[15,19)], but the products obtained are amorphous up to 350 °C[15,120)]. Crystallization of the MoS_2 produced begins at this temperature as already indicated, but thermal degradation of MoS_3 yields only the hexagonal form of MoS_2 [15)] rather than the rhombohedral[19)]. The latter conclusion was reached[19)] on misinterpretation of the X-ray powder diagrams[15)]. Heated MoS_3 yields hexagonal MoS_2 with considerable stacking faults[15)]. Thermogravimetric analysis of MoS_3 produced by the decomposition of $(NH_4)_2MoS_4$ in the presence of air showed that oxidation of the trisulfide begins at 200 °C[7)].

Molybdenum Pentasulfide

The preparation of this compound was reported in 1916[122)], but no recent reports have appeared in the literature. The preparation involves the reduction with zinc of an ammonium molybdate solution in 10% H_2SO_4 and precipitation of $Mo_2S_5 \cdot 3\,H_2O$ with hydrogen sulfide. The solid was described as brown which, upon careful heating in carbon dioxide, gave the black solid Mo_2S_5. Efforts in this laboratory[7)] to reproduce it by the method described in the literature yielded materials containing variable S/Mo atomic ratios. The solids thus obtained were amorphous to X-rays and their infrared spectra showed the presence of Mo-O bands other than those due to water. It therefore appears that the preparation of pure Mo_2S_5 is still in doubt.

Other Molybdenum Sulfides

A tetrasulfide of molybdenum, MoS_4, was first reported by Berzelius[6,123)], but efforts to reproduce this failed[7)]. No monosulfide of molybdenum exists[8)], but a mixed molybdenum monosulfide, "stabilized" by tin, of composition Mo_6SnS_7 and specific gravity of 5.69, has been prepared as a black crystalline material by heating under vacuum a mixture of molybdenum, tin, and sulfur[8)]. No further work on this material has been reported. A new sulfide, Mo_3S_4, that is isostructural with Mo_3Se_4 has been prepared. Its rhombohedral structure consists of an Mo_6S_8 unit formed by a Mo_6 octahedral cluster inscribed in a deformed cube of sulfur atoms. The Mo-Mo distances in the cluster are 2.69 A and 2.86 A[222)].

Other Related Materials

Tungsten disulfide, like molybdenum disulfide, exists both in the hexagonal and rhombohedral forms[15)]. Thermal decomposition of WS_3 is similar to that of MoS_3.

The rhombohedral form is prepared from melt solutions[15]. W-S mixtures, in 1:2 ratio compressed at 45 kbars and heated at 1800 °C for 2 to 3 minutes, gave a silver-gray product which was completely rhombohedral[23]. No evidence was found for the existence of lower tungsten sulfides, such as W_2S_3 [15].

A three-layered rhombohedral form of $MoSe_2$ has been produced by subjecting the two-layered hexagonal form to pressures of 40 kbars at 1500 °C. This new form is isostructural with rhombohedral MoS_2 [124]. The phases previously regarded as Mo_2Se_3 and Mo_2Te_3 were found to correspond to the composition Mo_3X_4 and are better described as Mo_3Se_4 and Mo_3Te_4, respectively[111].

Mixed molybdenum chalcogenides like Mo_2S_3Se, Mo_2S_3Te, Mo_2Se_3S, Mo_2Se_3Te, Mo_2Te_3S, and Mo_2Te_3Se were synthesized by heating stoichiometric amounts of Mo_2X_3 and X^1 (X or X^1 = S, Se or Te) for 15 hours at 1000 °C in an evacuated quartz tube. The products are gray, crystalline and metallic in appearance. They have hexagonal symmetry and are isostructural with the corresponding MoX_2 [125].

Oxysulfides

The number of known metal oxysulfides is relatively small[126]. Three types have been well established; these are: ZrOS, La_2O_2S, and ThOS[126]. A variety of molybdenum oxysulfides have been postulated to exist. The existence of $MoOS_2$ and MoO_2S has been claimed to occur during the oxidation of MoS_2 [72,127–129]. The presence of molybdenum oxysulfides was also claimed[130] to be the result of the decomposition of sulfur-containing organic compounds of molybdenum. The oxysulfide MoOS was claimed to be present in sulfided molybdenum-containing catalysts[131], but it is now believed that this material may be a mixture of MoO_2 and MoS_2 [132].

The following three preparative routes to molybdenum oxysulfides have appeared in the literature:

1. Reaction of MoO_2Cl_2 with Na_2S in ethanol as given by ter Meulen[133]:

$$MoO_2Cl_2 + Na_2S \longrightarrow MoO_2S + 2\ NaCl$$

2. Reaction of MoS_2 and H_2O as described by Cannon[92,134]:

$$MoS_2 + H_2O(g) \longrightarrow MoOS_2 + H_2$$

3. Thermal decomposition of $(NH_4)_2MoO_2S_2$ as proposed by Spengler and Weber[130]:

$$(NH_4)_2MoO_2S_2 \xrightarrow{\Delta} MoOS_2 + 2\ NH_3 + H_2O$$

Work was carried out in this laboratory to reproduce these reactions[7]. Attempts to prepare MoO_2S did not yield a simple molybdenum oxysulfide or a molybdenum sulfide, but a mixture of unidentifiable products. The reaction of MoS_2 with water vapor yielded neither hydrogen nor hydrogen sulfide. The thermal decomposition of

$(NH_4)_2MoO_2S_2$ under nitrogen was studied by thermogravimetric analysis, and the results are shown in Fig. 12. The weight loss up to 150 °C corresponds to 18.5%; beyond this and up to 350 °C, there is a gradual loss up to 22.7%, the expected weight loss due to $NH_3 + H_2O$ being 22.8%. A sample of $(NH_4)_2MoO_2S_2$, when heated at 250 °C for two hours under helium, showed a weight loss of 20.7%. The product consisted of very small, shiny black needle-like crystals, but these were

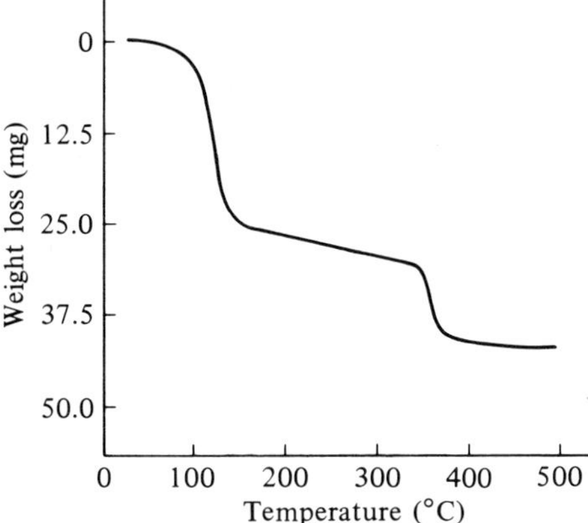

Fig. 12. The thermal decomposition of $(NH_4)_2MoO_2S_2$ under nitrogen (sample weight 134.9 mg)

amorphous to X-rays; the Mo : S ratio found was 1 : 1.93. Further heating up to 500 °C resulted in the evolution of sulfur dioxide. Although the chemical analysis does not differentiate between $MoOS_2$ and a 2 : 1 mixture of MoS_3 and MoO_3, the X-ray evidence rules out the presence of MoO_3.

The postulated existence of the oxysulfide MoOS can be ruled unlikely on the basis of the structures of MoO_2 und MoS_2. Molybdenum dioxide has a three-dimensional structure with distorted octahedral coordination (d^2sp^3 hybridization) about the molybdenum atoms[126] while MoS_2 has a layer structure with a trigonal prismatic coordination about the molybdenum atoms (d^4sp hydridization). Consequently, any existence of a MoOS species may be expected to be only transient.

Thiohalides

The literature on the thiohalides of molybdenum and tungsten is sparse and relatively recent. A red-brown solid of composition $Mo_5S_8Cl_9$ has been reported[135] which was prepared by the reaction of molybdenum metal with S_2Cl_2 in a carbon dioxide atmosphere. Tungsten metal similarly treated gave the dark red solid $W_2S_7Cl_8$ [135]. The tungsten-containing solid was found to be insoluble in water and alkalies and decomposed when sublimed. The product was not further characterized[135]. Other molybdenum thiohalides that were characterized only by analysis include $MoSCl_2$, which was isolated as an intermediate in the chlorination of MoS_2 with chlorine gas at 400 °C[136], and MoS_2Cl_3, a yellow-cinnamon powder obtained by the reaction of $MoCl_5$ with S_2Cl_2 at 250 °C in a sealed tube[137]. The solid, MoS_2Cl_3, slowly decomposed upon exposure to air and was found to be insoluble in alcohol, benzene, chloroform, and carbon tetrachloride. The solid was formulated as a dimer, $Mo_2S_4Cl_6$, analogous to molybdenum pentachloride, but no proof for this structure was provided[137].

Tab. 4. Thiohalides of molybdenum and tungsten

Molybdenum compounds	Ref.	Tungsten compounds	Ref.
$MoSCl_2$	136)	WS_2Cl_2	141)
$MoS_2Cl_3(Mo_2S_4Cl_6)$	137)	$WSCl_3$	142)
$MoSCl_3$	142)	$WSCl_4$	137, 142)
MoS_2Cl_2	138, 139, 141)	$WSBr_4$	142)
MoS_2Br_2	139)		
$Mo_2S_4Cl_5$	139)		
$Mo_2S_5Cl_3$	139)		
$Mo_2S_5Br_3$	139)		
$Mo_2S_4OCl_2$	138)		
$Mo_3S_7Cl_4$	144)		
$Mo_3Se_7Cl_4$	144)		

The reported thiohalides of molybdenum and tungsten are given in Table 4.

The reaction of metals and their oxides with disulfurdichloride has been investigated[138, 139]. Molybdenum metal reacts with S_2Cl_2 at 475 °C to yield a dark brown sublimate which, after extraction with carbon disulfide, analyzed as MoS_2Cl_2 [138]. Similarly, the compounds MoS_2Cl_2 and MoS_2Br_2 (X = Cl or Br) in 1 : 1 molar ratio at 500–525 °C[139]. These materials are brown, hydrolyze slowly when exposed to the air, and decompose in a nitrogen atmosphere above 500 °C to MoS_2 and the free halogen. X-ray powder diagrams indicate that the chloride and bromide are isomorphous[139]. When Mo metal is heated with excess S_2X_2 (X = Cl or Br) at temperatures lower than 450 °C, the species $Mo_2S_5Cl_3$ and $Mo_2S_5Br_3$ result[139].

The MoS_2X_2 and $Mo_2S_5X_3$ species prepared are diamagnetic; the oxidation state of molybdenum is believed to be +4 in analogy with the corresponding NbS_2X_2 [139]. The vibrational spectra of $Mo_2S_5X_3$ (X = Cl or Br) suggest that they contain S_2 groups that bridge two Mo atoms, similar to those in NbS_2X_2 rather than those in pyrites types[140]. The MoS_5Br_3 compound can also be prepared by the reaction of sulfur with $MoBr_2$ [140].

Molybdenum trisulfide reacts with S_2Cl_2 at 350–400 °C to give the brown solid $Mo_2S_4Cl_5$, and with S_2Cl_2 or S_2Br_2 at 420–480 °C to yield the red solids $Mo_2S_5Cl_3$ or $Mo_2S_5Br_5$, respectively[139]. They are stable in air; their X-ray powder patterns have been given. Thermogravimetric analysis in nitrogen shows that the thiohalides are stable up to 370 °C, beyond which they decompose to yield MoS_2. Molybdenum trioxide is formed when the thiohalides are heated in oxygen at 250 °C [139].

Molybdenum and tungsten oxytetrachlorides, $MoOCl_4$ and $WOCl_4$, react with anhydrous H_2S in boiling benzene to form the brown solids MoS_2Cl_2 and WS_2Cl_2[141].

The synthesis of the thiohalides $MoSCl_3$, $WSCl_3$, and WSX_4 (X = Cl or Br) has been reported[142]. The compound $WSCl_4$ is obtained in 70% yield by the reaction of stoichiometric quantities of WCl_6 and Sb_2S_3[142]. The reaction of sulfur with WCl_6[137,142] or WCl_5[142] gives in 100% yield $WSCl_4$ (mp 142–146 °C). Even when excess sulfur is used, $WSCl_4$ is the only thiochloride formed[142]. The reaction of Sb_2S_3 at 150 °C with the pentachlorides $MoCl_5$ and WCl_5 gives the thiohalide $MSCl_3$. Attempts to prepare $WSCl_3$ by the reduction of $WSCl_4$ with aluminum were not successful. All of these thiohalides are unstable in air and evolve hydrogen sulfide and hydrogen halide. $WSCl_4$ sublimes in vacuum to yield diamagnetic[142] ruby-red crystals[137, 142], whereas $WSBr_4$ volatilizes at 180–200 °C to give dark green crystals[142]. The thiochlorides $MoSCl_3$ (greenish-black) and $WSCl_3$ (black) are nonvolatile[142]. Low magnetic moments (0.75 BM for $MoSCl_3$ and 0.54 BM for $WSCl_3$) are attributed to interactions of electrons on adjacent metal atoms through a non-linear M-S-M systems[142]. The absence of infrared peaks above 383 cm^{-1} in the spectra of $MSCl_3$ is consistent with polymeric structure with M-S-M bridging[140,142]. Bands at 569 cm^{-1} for $WSCl_4$ and 555 cm^{-1} for $WSBr_4$ have been assigned to terminal W=S bonds[142]. A recent crystal structure determination of $WSCl_4$ and $WSBr_4$ has been carried out[143]. The structure of $WSCl_4$ (see Fig. 13) shows that the arrangement of the five ligands around the tungsten atom is square pyramidal, where two molecules of $WSCl_4$ associate through the weak W-Cl bonds across a center of symmetry to form $W_2S_2X_8$ dimeric units[143].

The reaction of molybdenum trichloride with sulfur or selenium at 450 °C in a sealed tube was reported to give, in 20–24 hours, $Mo_3S_7Cl_4$ or $Mo_3Se_7Cl_4$, respectively[144]. The X-ray powder diagrams of these materials were obtained. The sulfur compounds decompose at 530 °C to give MoS_2, whereas the selenium analog

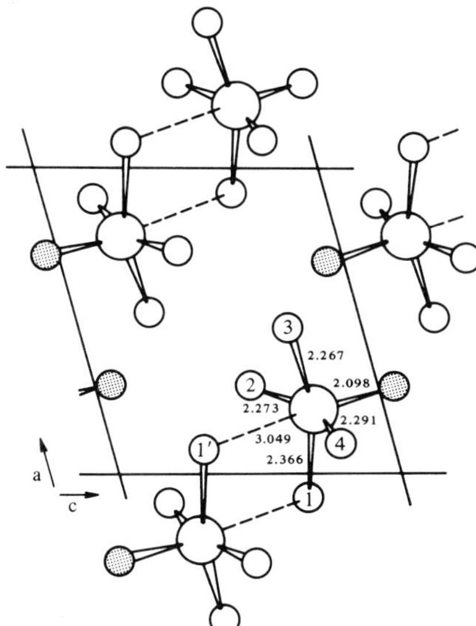

Fig. 13. The contents of the unit cell of $WSCl_4$ in the b projection (large open circles, tungsten; small open circles, chlorine; small closed circles, sulphur)

gives $MoSe_2$ at 450 °C. The same decomposition products are obtained when the parent compounds are reduced with hydrogen gas at 285 and 265 °C, respectively[144].

The product, $Mo_5S_8Cl_9$, obtained by the reaction of Mo metal and S_2Cl_2[135] when washed with boiling water gave a solid which was shown by analysis to be $Mo_2S_4OCl_2$[138]. The same compound, $Mo_2S_4OCl_2$, resulted when MoO_3 was reacted with S_2Cl_2 at 350 °C for 8 hours, the resulting amorphous yellow-brown solid being washed with carbon disulfide before analysis. This oxythiochloride of molybdenum was not further characterized[138]. The reaction of tungsten metal with S_2Cl_2 in a nitrogen atmosphere was found to yield only WCl_6 and free sulfur rather than $W_2S_7Cl_8$[135]. When the same reaction was carried out in the presence of air, $WOCl_4$ formed[138]. Tungsten trioxide reacted with S_2Cl_2 to yield $WOCl_4$ rather than a tungsten oxythiochloride[138].

Thiomolybdites

The thiomolybdites are a class of molybdenum-sulfur compounds which contain molybdenum in a low oxidation state, usually +3. Two main types of such materials exist. The first type has the formula $MMoS_2$ where M is a monovalent cation, usually an alkali metal. The second type has the formula MMo_2S_4 where M is a divalent cation, usually a transition metal. There are other thiomolybdite species, of composition other than that described above, which have been identified in ternary phase studies involving the M-Mo-S system (M = a transition element), but these have not been well characterized.

Thiomolybdites of Type $MMoS_2$

The sodium, potassium, rubidium, and cesium thiomolybdites, namely $NaMoS_2$ (brilliant black), $KMoS_2$, $RbMoS_2$ and $CsMoS_2$ (all black with a metallic lustre), were prepared by the reaction of the corresponding alkali metal sulfide with Mo_2S_3[145]. They all have identical X-ray powder patterns. The potassium, rubidium, and cesium compounds undergo decomposition with the evolution of sulfur when heated at 320, 380 and 420 °C, respectively[145].

When the alkali metal molybdates or tungstates are treated with CS_2 or H_2S vapors at 300–320 °C, materials of composition M_2Mo(or W)S_4 result, but so far they have not been further characterized[145]. The sodium, potassium, and rubidium thiotungstates thus prepared, when reduced by hydrogen, yield the thiotungstites $NaWS_2$, KWS_2, and $RbWS_2$ which are black solids having a metallic lustre. They are stable up to their fusion point. Unlike the corresponding thiomolybdites, these cannot be prepared from the alkali sulfide and W_2S_3[145] since the latter does not exist[15]. The $RbWS_2$ cannot be prepared free of Rb_2WS_4, but the latter can be removed by washing with water[145]. The lithium and cesium thiotungstites could not be prepared. The X-ray patterns of MWS_2 are identical with those of the corresponding molybdenum compounds.

Lithium thiomolybdite, $LiMoS_2$, has been prepared by the hydrogen reduction of $Li_2MoS_{3.5}$ at 600 °C. The latter is obtained by the reduction of lithium thiomolybdate[145]. Reduction of K_2MoS_4 with hydrogen at 650 °C gives the homogeneous solids $K_6Mo_3S_8$ with formation of hydrogen sulfide[146]. These are black metallic-looking solids (density = 2.63 and 3.47 g/cm^3, respectively) which are unstable in air. The presence of $KMoS_2$ could not be detected during the reduction of K_2MoS_4 with hydrogen[146]. A thallous thiomolybdite of composition $TlMoS_2$ has been described[147] and was reportedly prepared by the addition of H_2S to solutions of molybdate and thallous ion. The solid is amorphous to X-rays, but no other data are given. It is unlikely that a thiomolybite could be produced under these conditions.

Thiomolybdites of Type MMo_2S_4

Thiomolybdites of divalent transiton elements having the formula MMo_2S_4 (M = Ti, V, Cr, Mn, Fe, Co, Ni) were prepared recently for the first time[148]. Their preparation consisted of heating mixtures of M + 2 Mo + 4 S or M + 2 MoS_2 in a silica tube under vacuum at 1100 °C for periods of 20 hours. The nickel compound could be prepared only if reactive nickel [obtained from the decomposition of $Ni(CO)_4$] was used. These materials are black crystalline solids that are stable in air. Their X-ray powder diagrams have been given[148].

The solid $CoMo_2S_4$ was also prepared by heating the elements at 1100 °C in an evacuated silica tube[149]. The solid has one unpaired electron localized on the cobalt atom, the formula of the solid thus being $Co^{+2}Mo_2{}^{+3}S_4$. X-ray structure determination has shown[149,150] that the Co and Mo atoms are within distorted sulfur octahedra, the Mo-Mo distance being 2.85 Å, indicative of metal-metal bonding[149]. Crystallographic parameters for the monoclinic $FeMo_2S_4$ and $CoMo_2S_4$ have also been reported[151]. Recent X-ray diffraction on crystal of $CoMo_2S_4$ and $FeMo_2S_4$ has shown these to be monoclinic and their structures to consist of octahedral chains of MoS_6 and CoS_6 or FeS_6 extending along the a and b axes[152].

The phases of the chromium-molybdenum-sulfur system have been examined in detail[153]. The solids found are: monoclinic $CrMo_2S_4$, and triclinic $CrMo_2S_2$ and $CrMo_3S_4$. The solids have a magnetic moment of 4.9 BM; the chromium is postulated to be present as high spin Cr^{+2}, the molybdenum not contributing materially to the magnetic moment due to Mo-Mo bonding[153] as found in $CoMo_2S_4$[149].

Other Metal Thiomolybdites

The preparation of metal thiomolybdites or thiotungstites of type AMS_y, where A is Fe, Co or Ni; M is Mo or W; and y is 2.5 to 3, was carried out by heating mixtures of the metal sulfides of A and M in the ratio of 1 : 2 from 500–1300 °C in an inert atmosphere[154]. The metal sulfides were also prepared *in situ* from the metals or their oxides by treatment with H_2S. Thus, the solids $FeMoS_{2.5}$, $FeWS_3$, $NiMoS_3$, and $CoMoS_3$ were obtained. They were shown to be new compounds by X-ray powder pattern analysis. The iron compounds are ferromagnetic. The tungsten solid is gray-black, whereas the molybdenum compounds are steel-gray metallic crystals. These products were suggested for use as ceramic pigments[154)].

The preparation of new sulfides of formula $M^{+2}Mo_nS_{n+1}$ has been reported[155)]. The compounds are stoichiometric when M is Ag, Sn, Ca, Sr, Pb, or Ba. If M is Ni, Co, Fe, Cr, Mn, Cu, Mg, Zn, Cd, solid solutions are observed with $2 \leqslant n \leqslant 6$. The series of formula $M_2Mo_nS_{n+1}$ was also prepared where M is an alkali metal and n is 2 or 5. These phases obtained are generally rhombohedral with an eventual triclinic distortion[155)].

The ternary system Cu–Mo-S has also been investigated at 800 °C. The presence of the compound $CuMo_2S_3$ was ascertained[156)]. Data obtained on the reaction of Cu with MoS_2 at ~750 °C indicate that the products formed are Cu_2S and $CuMo_2S_3$[7)]. The ternary system Fe–Mo-S has also been studied, and the compound Mo_5FeS_{11} stable below 600 °C was found[157)].

No further characterization of these materials is available at present.

Reaction of Molybdenum Compounds with H_2S/H_2

Current emphasis in hydrodesulfurization of fuels has generated considerable interest in the reactions of molybdenum compounds with hydrogen sulfide or mixtures of H_2S/H_2. The following section will outline some of the pertinent reactions of molybdenum compounds with hydrogen sulfide or mixtures of H_2S/H_2. Details of the chemistry of hydrodesulfurization catalysts containing molybdenum are referenced later in this report.

Equilibria in the M–S–H_2 System

The various sulfides of molybdenum have already been discussed. The equilibria which exist between the lower sulfides of molybdenum and mixtures of hydrogen and hydrogen sulfide between 850 and 1200 °C have been studied by Stubbles and Richardson[116)] using a radiochemical method. The results obtained, shown in Fig. 14, indicate that Mo_2S_3 does not exist below 600 °C under the conditions described and, therefore, it is doubtful that it is present in sulfided hydrodesulfurization catalysts. As pointed out earlier, these results are in agreement with those obtained by McCabe[65)] concerning the presence of Mo_2S_3 in equilibria of this type and contrary to those obtained by others[64)] denying the existence of Mo_2S_3.

The reaction of powdered molybdenum or tungsten with hydrogen sulfide has been examined. These metals, when treated at low pressures and at 640–700 °C and 350–450 °C, respectively, yield MoS_2 and WS_2 with a lamellar structure[158)].

Reactions of MoO_3 with H_2S/H_2

The reaction of molybdenum trioxide and lower oxides with H_2 and/or H_2S has been reported. Romanowski studied[13)] the reaction of MoO_3, Mo_4O_{11} and MoO_2

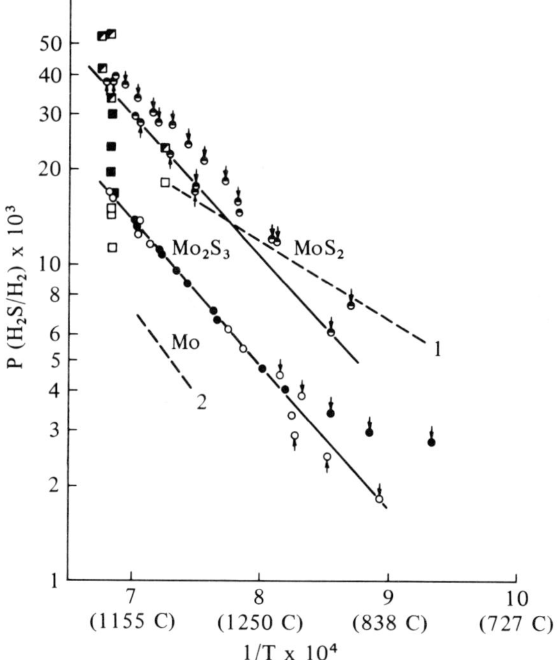

Fig. 14. Equilibria involving Mo + Mo_2S_3 and Mo_2S_3 + MoS_2 between 700 and 1200 °C
Full Lines: Stubbles & Richardson's results[17]; Broken Lines: (1) Parravano & Malquori, allegedly Mo + MoS_2; (2) McCabe, Mo + Mo_2S_3. Equilibria: ◓, Mo_2S_3 + MoS_2; ◒, Mo + Mo_2S_3 (30 wt%S in mixture); ○, Mo + Mo_2S_3 (9 wt%S). Equilibrium phases identified; ◩, MoS_2; ■, Mo_2S_3; □, Mo. Arrows indicate whether points were obtained after increase ↑ or decrease ↓ in temperature, *i.e.*, from low or high H_2S pressures

with hydrogen sulfide at 400–500 °C and found by chemical and X-ray analysis that MoO_2 and MoS_2 were the only reduction products. Richardson, however, reported that 10% MoO_3 on γ-Al_2O_3 sulfides completely in 2% H_2S in H_2 at 400 °C to yield MoS_2[132]. The reaction of MoO_3 in H_2S has been recently examined in detail by Seshadri, Massoth and Petrakis[14]. Their data were obtained by electron spin resonance and microbalance studies. Sulfiding of bulk MoO_3 with H_2S/H_2 at 300–500 °C results in the presence of MoO_2 and MoS_2 only, as found by others[13], but further sulfiding of MoO_2 to MoS_2 is very slow. Molybdenum trioxide on silica behaves like bulk MoO_3[13]. The sulfiding of MoO_3 on alumina, however, proceeds through the formation of a pentavalent form of Mo in addition to MoS_2 and MoO_2. Sulfiding at lower temperatures favors the formation of Mo(V) and MoS_2, whereas at higher temperatures, MoO_2 is formed in addition to MoS_2 and Mo(V)[14].

Reactions of Molybdates with H_2S/H_2

Major emphasis on the nature of the reduction products of molybdates with H_2S/H_2 mixtures has been carried out on cobalt molybdate, which is the standard hydrodesulfurization catalyst. At present, however, there is still no definitive evidence indicating the exact species present on sulfided "cobalt molybdate" catalysts.

The reduction of unsupported stoichiometric cobalt molybdate was studied up to 1000 °C using dry H_2S gas[159,160]. Reduction begins above 400 °C, the products containing both oxygen and sulfur up to 750 °C. Above that temperature, materials of composition $CoMoS_{3.13}$ are indicated. At 1000 °C, the products are mixtures of MoS_2 and CoS[159,160].

A magnetic study of cobalt molybdate catalysts has been carried out[132]. The results obtained showed that when pure cobalt molybdate is sulfided in 2% H_2S in H_2, a mixture of Co_9S_8, MoO_2, and MoS_2 results. The alumina-supported catalyst under hydrodesulfurization conditions consists of Al_2O_3, $CoAl_2O_4$, Co_9S_8, MoS_2, and MoO_2[132]. In recent work, Mitchell[161] indicated that sulfided cobalt molybdate catalysts do not contain discrete sulfides of molybdenum and cobalt; molybdenum is only partially sulfided, and tetrahedral cobalt or molybdenum are sulfided in preference to octahedral species.

It is hoped that further work in this area carried out with modern analytical techniques will shed more light on the nature of the species present in sulfided molybdate catalysts.

Thiomolybdates and Thiotungstates

The thiomolybdates and thiotungstates constitute an important area of inorganic molybdenum-sulfur chemistry; considerable elucidation has taken place only in recent years. Although the thioanions MoO_3S^{-2}, $MoO_2S_2^{-2}$, $MoOS_3^{-2}$, MoS_4^{-2}, WO_3S^{-2}, $WO_2S_2^{-2}$, WOS_3^{-2}, and WS_4^{-2} have been described in the older literature[162,164], only the preparation of the ammonium and alkali salts of MoS_4^{-2} and WS_4^{-2} could be reproduced in recent work[165]. Attempts to prepare the $MoOS_3^{-2}$ anion according to procedures described in recent literature[166], also failed[166], although the corresponding WOS_3^{-2} anion was reproduced[165] as described[166]. The study of the chemistry of the thiomolybdate and thiotungstate anions has been primarily carried out by Müller and co-workers at Göttingen and is described below.

Equilibria in Solution

Thiomolybdate and thiotungstate anions are prepared from aqueous solutions by the reaction of MoO_4^{-2} or WO_4^{-2} with hydrogen sulfide. During this reaction, the solution assumes a yellow color which later becomes orange and finally red, denoting the increasing substitution of oxygen by sulfur in the molybdate or tungstate anion. The relative ease of isomorphous substitution of oxygen by sulfur in the tetrahedrally coordinated thiomolybdate or thiotungstate anions, unlike the molybdenum oxysulfides discussed earlier, is possible here since no condensation of these anions occurs[126]. Condensation, which might be present in oxysulfides, leads to layer structures with those of oxygen being entirely different than those of sulfur[126].

The equilibria that exist in thiomolybdate solutions have been studied by several workers. The systems Na_2MoO_4–Na_2S–NaOH–H_2O and Na_2WO_4–Na_2S–NaOH–H_2O were studied[167]. In this work, Na_2WS_4 is claimed to form only in 18% NaOH, whereas Na_2MoS_4 is reported to be stable in solutions greater than 27% NaOH. In more dilute sodium hydroxide, both compounds undergo hydrolysis, even in a large excess of sulfide ion, but neither Na_2MoS_4 or Na_2WS_4 could be isolated in pure form[167]. A method for preparing Na_2MoO_3S has been described in the literature[168], but efforts to reproduce it were not successful[7]. A spectrophotometric investigation of equilibria present in MoO_4^{-2}/H_2S solutions has shown the existence of the $MoO_2S_2^{-2}$, $MoOS_3^{-2}$ and MoS_4^{-2} anions, but the authors failed to consider the presence of MoO_3S^{-2} in these solutions in the interpretation of their spectra[169,170]. However, Müller and co-workers in reexamination of the electronic spectra of solutions of MoO_4^{-2}/H_2S and WO_4^{-2}/H_2S have provided evidence for the existence of MoO_3S^{-2} and WO_3S^{-2} [158]. The spectra for the MoO_4^{-2}/H_2S system are shown in Fig. 15. The absorption maxima for both MoO_4^{-2}/H_2S and WO_4^{-2}/H_2S are given in Table 5. The spectra shown in Fig. 15

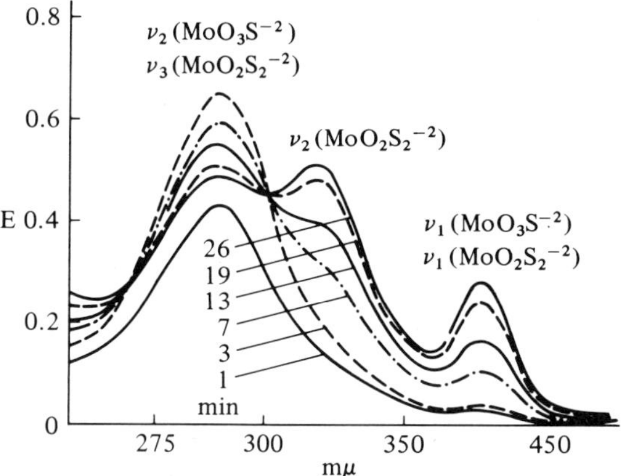

Fig. 15. Electronic spectra of the reaction mixture containing 1×10^{-4} M MoO_4^{-2} and 10 ml saturated solution of H_2S in aqueous solution showing formation of MoO_3S^{-2} and $MoO_2S_2^{-2}$ ions

were taken at different reaction stages. The band at 288 mμ is first seen in the spectrum two minutes after mixing H_2S and MoO_4^{-2}; it gains intensity quickly and reaches a maximum in about five minutes. Simultaneously, a weak band appears at 392 mμ. A shoulder gradually develops at 319 mμ along with intensification of the 392 mμ band and reduction in the intensity of the 288 mμ band. The 288 and 392 mμ bands are attributed to the MoO_3S^{-2} species[171]. Similar experiments have shown that the WO_3^{-2} anion has absorption maxima at 327 and 244 mμ. Unlike the MoO_4^{-2} ion which forms tri- and tetrathiospecies readily, the reaction of the WO_4^{-2} ion under the same experimental conditions does not proceed beyond the $WO_2S_2^{-2}$ stage even after twelve hours[171]. All thiomolybdate anions in solutions of pH lower than 7 give molybdenum trisulfide[170].

Tab. 5. Electronic spectra of different thioanions of molybdenum and tungsten[1])

Mono	Di	Tri	Tetra
		Thiomolybdates	
ν_2 = 288	ν_3 = 288 (0.30)	ν_5 = 243 (1.40)	ν_3 = 241 (2.95)
ν_1 = 392	ν_2 = 319 (0.60)	ν_4 = 270	ν_2 = 316 (1.80)
	ν_1 = 394 (0.30)	ν_3 = 319 (0.66)	ν_1 = 463 (1.30)
		ν_2 = 392 (0.87)	
		ν_1 = 460 (0.23)	
		Thiotungstates	
ν_2 = 244	ν_3 = 244 (0.39)	ν_4 = 243 (0.97)	ν_2 = 276 (2.85)
ν_1 = 327	ν_2 = 273 (0.69)	ν_3 = 270 (0.72)	ν_1 = 391 (1.85)
	ν_1 = 327 (0.40)	ν_2 = 330 (1.13)	
		ν_1 = 380 (0.3)	

[1]) $\nu = \lambda_{max}$ in mμ.
$\epsilon_{values} \times 10^4$ in parentheses (in $1 \cdot mol^{-1} \cdot cm^{-1}$).

The existence of the MoO_3S^{-2} and WO_3S^{-2} anions in molten sodium fluoride has been shown by cryoscopy. Thus MoO_3 or WO_3 react with sodium sulfide in sodium fluoride to give $NaMoO_3S$ and $NaWO_3S$, respectively, which are stable up to 1000 °C[172]).

Spectrophotometric evidence for the existence of MoO_3Se^{-2} and WO_3Se^{-2} in aqueous solution has also been obtained. The spectra were obtained on solutions of sodium molybdate or tungstate in which H_2Se was introduced. Evidence for the existence of these anions has been presented for the first time[173]). The preparation of the sulfur-bridged anion $Mo_2S_2O^{+2}$ has been reported[174]). This ion was found to be stable in 10 M HCl solution for several weeks.

Preparation and Properties of Solids

Monothio Anions

The monothio anions MoO_3S^{-2} and WO_3S^{-2} have been shown so far to exist only in solution, and it has not yet been possible to isolate pure monothio compounds[171]).

As already pointed out, efforts to prepare the salt, Na_2MoO_3S, as described in the literature[168] were unsuccessful[7]. No salts of the selenium analogs MoO_3Se^{-2} and WO_3S^{-2} have been isolated[173]. The related monothio anions ReO_3S^- and TcO_3S^- have been prepared in solution[175], but only the thallous salt of the ReO_3S^- anion, $TlReO_3S$, has been isolated[176].

Dithio Anions

The preparation and properties of various salts of the $MoO_2S_2^{-2}$ and $WO_2S_2^{-2}$ anions have been described[177]. The salts $(NH_4)_2MoO_2S_2$, $(NH_4)_2WO_2S_2$, and $K_2MoO_2S_2$ are prepared[177] according to methods established in the old literature[163,164] by the passage of hydrogen sulfide into cold solutions of molybdate or tungstate anion containing the desired cation. The alkali salts are very soluble and diffficult to isolate[177]. Addition of cesium chloride to solutions of $(NH_4)_2MoO_2S_2$ or $(NH_4)_2WO_2S_2$ gives the trithio salts $CsMoOS_3$ and Cs_2WOS_3 rather than the dithio cesium salt[177]. The formation of the trithio salt has been attributed to the lower solubility of the cesium salt of the $MoOS_3^{-2}$ anion. The latter anion results in solution from the hydrolysis of $MoO_2S_2^{-2}$ which yields H_2S which then reacts with $MoO_2S_2^{-2}$ to yield $MoOS_3^{-2}$ according to the reaction[177]:

$$MoO_2S_2^{-2} + H_2S \longrightarrow MoOS_3^{-2} + H_2O$$

Only salts of the dithio anions with large cations can be prepared[177,178]. Addition of Co^{+2}, Pb^{+2}, and Ce^{+3} to solutions of ammonium dithiomolybdate yielded solids that showed the absence of Mo-S bonding at 480–450 cm^{-1} [178]. Dithio salts so far prepared[177] are given in Table 6. The salt $Cs_2WO_2Se_2$ has been also isolated as a stable solid by the addition of cesium chloride to a water solution of $(NH_4)_2WO_2S_2$ [177].

Tab. 6. Salts of $MoO_2S_2^{-2}$ and $WO_2S_2^{-2}$ anions

Salt	Color	Stability
$(NH_4)_2MoO_2S_2$	Orange	Stable in air
$K_2MoO_2S_2$	Orange	Stable in air
$(NH_4)_2WO_2S_2$	Yellow	Stable in air
$Ni(NH_3)_6MoO_2S_2$	Yellow	Gives off NH_3 slowly upon exposure of air; black solid remains behind[1])
$Ni(NH_3)_6WO_2S_2$	Light yellow	Gives off NH_3 slowly upon exposure of air; black solid remains behind[1])
$Tl_2MoO_2S_2$	Orange	Stable in air
$Tl_2WO_2S_2$	Yellow	Stable in air

[1]) Black solid shows absence of infrared band at 480–450 cm^{-1} due to Mo-S bonding (Ref.[193]).

The salts $(NH_4)_2MoO_2S_2$ and $(NH_4)_2WO_2S_2$ are soluble in water, liquid ammonia, dimethylsulfoxide, dimethylformamide, and morpholine but are practically insoluble in ethyl alcohol, ether, and carbon disulfide[177)].

Trithio Anions

Although the preparation of dithiomolybdates and tetrathiomolybdates by the introduction of H_2S into molybdate solutions presents no difficulties, the preparation of crystalline trithiomolybdates by this route has only been recently accomplished[179)]. Thus passage of H_2S into 1 : 1 NH_3 solution containing ammonium molybdate yielded, after the addition of excess aqueous CsCl solution and acetic acid (up to pH 10), orange to red crystals of Cs_2MoOS_3. The less soluble Tl_2MoOS_3 was also prepared by the addition of thallous nitrate to cesium trithiomolybdate solutions[179)]. The cesium salt gives a neutral reaction in solution. The equivalent conductivity of the $MoOS_3^{-2}$ ion is $\Lambda°_{25} = 77$ ohm^{-1} cm^2 equiv^{-1}, that for MoO_4^{-2} being 74.5 ohm^{-1} cm^2 eguiv^{-1} [179)]. Another technique for preparing insoluble salts of the trithio anions involves the addition of the desired cation to a solution of $MoO_2S_2^{-2}$ or $WO_2S_2^{-2}$ which yields the $MoOS_3^{-2}$ or WOS_3^{-2} anions upon hydrolysis, as already shown[166,180)]. Thus the red $[(CH_3)_4N]_2MoOS_3$ and the yellow $[(CH_3)_4N]_2WOS_3$ salts as well as the corresponding tetraethylammonium derivatives have been isolated[166)]. Metal salts of trithio anions isolated[180)] are given in Table 7.

Tab. 7. Salts of $MoOS_3^{-2}$ and WOS_3^{-2} anions

Salt	Color	Stability
Cs_2MoOS_3	Orange	Stable in air
Cs_2WOS_3	Yellow	Stable in air
Tl_2MoOS_3	Red-orange	Stable in air
Tl_2WOS_3	Dark yellow	Stable in air
$[Ni(NH_3)_6]MoOS_3$	Orange-yellow	Stable in dry air
$[Ni(NH_3)_6]WOS_3$	Green-yellow	Stable in dry air

The salts Cs_2MoOS_3 and Cs_2WOS_3 are soluble in water to give orange-red and yellow solutions, respectively. Both salts are insoluble in alcohol or ether. The aqueous solutions of these salts hydrolyze slowly to give the dithio anions[180)].

Dithioselenomolybdate and dithioselenotungstate anions. $MoOS_2Se^{-2}$ and WOS_2Se^{-2}, have been prepared according to the reactions:

$$MoO_2S_2^{-2} + H_2Se \longrightarrow MoOS_2Se^{-2} + H_2O$$

$$WO_2S_2^{-2} + H_2Se \longrightarrow WOS_2Se^{-2} + H_2O$$

and the cesium, thallous and $Ni(NH_3)_6^{+2}$ salts of these anions have been isolated[181)]. The cesium salts are soluble in water; the molybdate giving an orange-red solution, the tungstate a yellow solution. The thallium and $Ni(NH_3)_6^{+2}$ salts are insoluble. All of these salts are practically insoluble in alcohol and ether. Water solutions of these anions form, upon long standing, red selenium and a brown precipitate. The color and stability of the isolated salts prepared are given in Table 8.

Tab. 8. Salts of the $MoOS_2Se^{-2}$ and WOS_2Se^{-2} anions

Salt	Color	Stability
Cs_2MoOS_2Se	Deep red	Stable in dry nitrogen atmosphere
Tl_2MoOS_2Se	Red-brown	Slow decomposition giving a black color
$[Ni(NH_3)_6]MoOS_2Se$	Red	Gradual decomposition with evolution of ammonia and black color formation
Cs_2WOS_2Se	Yellow-orange	Relatively stable in dry air
Tl_2WOS_2Se	Red-orange	Stable in dry nitrogen atmosphere
$[Ni(NH_3)_6WOS_2Se$	Yellow	Gradual decomposition with evolution of ammonia and black color formation

The preparation and properties of K_2MoOS_3 and K_2WOS_3 have been reported[182)], as well as the preparation of the K^+, NH_4^+, Rb^+, and Cs^+ salts of the anions $MoOSe_3^{-2}$, $WOSe_3^{-2}$, $MoOSSe_2^{-2}$, and $WOSSe_2^{-2}$ [183)]. The compounds $K_3(MoOS_3)Cl$ and $K_3(MoOS_3)Br$ have also been prepared and characterized[184)].

Tetrathio Anions

The tetrathiomolybdate, MoS_4^{-2}, and tetrathiotungstate, WS_4^{-2}, anions have been known for a long time. The ammonium salts of these are prepared by passing H_2S gas into ammonium molybdate or tungstate solutions[162, 164, 181)]. Although these and the potassium salts are readily isolated, the so-called tetrathiomolybdates of Co(II), Ni(II), Cu(II), and Zn(II) have been shown by X-ray powder diffraction, infrared and diffuse reflectance spectra to be mixtures of the sulfides of these metals and molybdenum trisulfide[185)]. Although $PbMoS_4$ has been claimed to exist[186)], it has also been found to be a mixture of PbS and MoS_3[178)]. The reaction of Fe^{+2}, Mn^{+2}, and UO_2^{+2}, and UO_2^{+2} also leads to decomposition of the MoS_4^{-2} anion[178)]. Apparently, the size of the cation determines the stability of the MoS_4^{-2} salts. For example, the salts $[Ni(NH_3)_6]MoS_4$ and $[Co(NH_3)_6]MoS_4$ have been isolated[165, 180)]. The salts of the MoS_4^{-2} and WS_4^{-2} anions and their selenium analogs[183)] which have been isolated are given in Table 9 along with their properties.

The salts containing the organic cations have been prepared by Leroy[166)], the metal ammine salts by Müller[180)], and those of selenium also by Müller[183)]. The decomposition of the metal ammine salts has been described as proceeding according to the equation[165, 180)]:

$$[Ni(NH_3)_6]MS_4 \longrightarrow 6NH_3 + NiS + MS_3 \ (M = Mo \text{ or } W)$$

Organic salts of the MoS_4^{-2} anion, such as tri-n-butylammonium and cyclohexylammonium, are extractable in organic solvents[187]. The free acids H_2MoS_4 and H_2WS_4 have been prepared by the reaction of $(NH_4)_2MoS_4$ or $(NH_4)_2WS_4$ and HCl in dimethylether at –78°C. The tungsten compound is the more stable. It decomposes directly into WS_3 and H_2S[188]. Aqueous solutions of MoS_4^{-2} and WS_4^{-2} upon acidification yield the insoluble trisulfides. It has been reported that the acidification

Tab. 9. Salts of the MoS_4^{-2}, WS_4^{-2}, $MoSe_4^{-2}$ and WSe_4^{-2} anions

Salt	Color	Stability
$(NH_4)_2MoS_4$	Red	Stable in dry air
K_2MoS_4	Red	Stable in dry air
$[Ni(NH_3)_6]MoS_4$	Orange-red	Moderately stable in dry air
$[Co(NH_3)_6]MoS_4$	Yellow-red	Decomposes with evolution of ammonia and formation of yellow color
$[(CH_3)_4N]_2MoS_4$	Red	Stable in dry air
$[C_2H_5)_4N_2MoS_4$	Red	Stable up to 180 °C in N_2
$(PH_4P)_2MoS_4$	Red	Stable in dry air
$(Ph_4P_2MoS_4$	Red	Stable in dry air
$(NH_4)_2WS_4$	Orange-yellow	Stable in dry air
K_2WS_4	Orange-yellow	Stable in dry air
$[Ni(NH_3)_6]WS_4$	Lemon-yellow	Stable in dry air
$[Co(NH_3)_6]WS_4$	Green-yellow	Slow decomposition with ammonia evolution and formation of black color
$[(CH_3)_4N]_2WS_4$	Orange-yellow	Stable in dry air
$[(C_2H_5)_4WS_4$	Yellow	Stable in dry air
$(Ph_4P)_2WS_4$	Yellow	Stable in dry air
$(Ph_4As)_2WS_4$	Yellow	Stable in dry air
$(NH_4)_2MoSe_4$		Stable in dry air
Rb_2MoSe_4	Blue	
$(NH_4)_2WSe_4$		
Cs_2WSe_4	Red	

of sodium thiomolybdate by the progressive addition of hydrochloric acid yields the polymerized species $Mo_4S_{15}^{-6}$, $Mo_2S_7^{-2}$, and $Mo_4S_{13}^{-2}$ at the pH ranges 7.5 to 8.5, 4.2 to 5.3, and 2.8 to 3.3, respectively; but below pH 2.5, the $MoS_3 \cdot H_2O$ species is formed[189]. In view of the difficulty in ascertaining the degree of polymerization of the more stable polymolybdates and polytungstates by such techniques[4,190], further work will be necessary to finalize these structures. Detailed preparation and properties of the NH_4^+, K^+, Rb^+, Cs^+, and Tl^+ salts of the MoS_4^{-2} and WS_4^{-2} anions have been presented recently[188].

Structures

In a single-crystal study, ammonium tetrathiomolybdate was shown to be isomorphous with β-K_2SO_4 and hence the MoS_4^{-2} ion is tetrahedral[191)]. A full structure determination of $(NH_4)_2$ WS_4 has shown it to consist of discrete undistorted tetrahedral WS_4^{-2} anions[192)]. The structures of $(NH_4)_2MoS_4$ and $(NH_4)_2WS_4$ have been shown to be identical[193)]. Recently, the crystal structure of cuprous ammonium tetrathiomolybdate, $CuNH_4MoS_4$, has been determined and is shown in Fig. 16[194)].

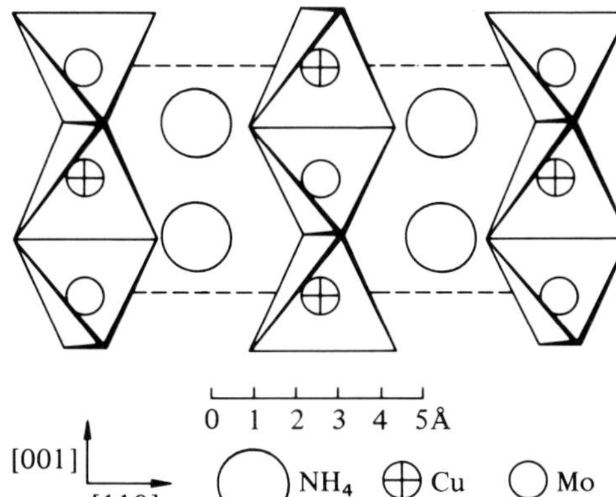

Fig. 16. A projection of the structure of $CuNH_4MoS_4$ along the [110] direction (sulfur atoms are located at the vertices of the tetrahedra.)

The structure consists of a tetrahedral arrangement of sulfur atoms around each molybdenum and copper atom, with each tetrahedron sharing two corners with each of its two neighbors. The ammonium ions are located in the spaces between the chains. The Mo-S distance is 2.19 Å in accord with those found in $(NH_4)_2MoS_4$[193)]. A structure determination has been carried out also on piperazinium tetrathiomolybdate, $(C_4H_{12}N_2)MoS_4$[195)]. The MoS_4^{-2} anion is tetrahedral with Mo-S distances 2.18 Å, and the piperazinium ion is in the chair form.

Recently, the preparation of salts of the bis(tetrathiotungsto)nickelate (II) ion, $Ni(WS_4)_2^{-2}$, has been reported and the tetraphenylphosphonium and -arsonium salts isolated. The structure shown in Fig. 17 has been proposed for this anion[196)].

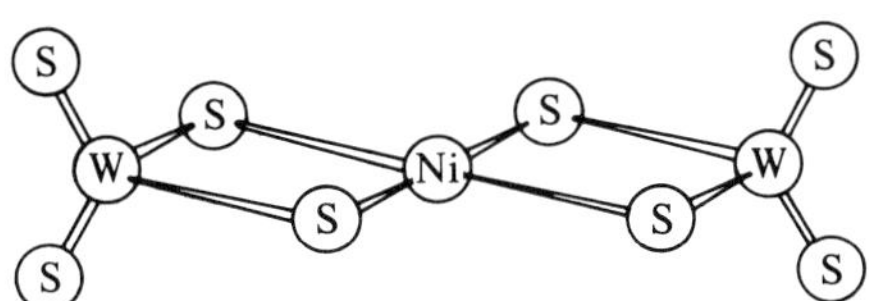

Fig. 17. Suggested structure for $[Ni(WS_4)_2]^{-2}$ (probable symmetry D_{2h})

X-ray data for various salts of MoS_4^{-2} and WS_4^{-2} have been given[188)]. X-ray crystallographic studies for dithio, trithio, and tetrathiomolybdates and -tungstates have also been reported[182,183,197)]. Similar data are scattered throughout several of the papers by Müller, which have been cited.

Thermal Stability

The thermal decomposition of $(NH_4)_2MoS_4$ and $(NH_4)_2MoO_2S_2$ has already been discussed under molybdenum sulfides and oxysulfides. Ammonium tetrathiomolybdate was found to decompose at $\approx$150 °C to yield MoS_3 which subsequently decomposes into MoS_2 [15]. The decomposition of $(NH_4)_2MoO_2S_2$ leads to a species of empirical composition $MoOS_2$, but no further proof of an actual oxysulfide was obtained[13]. The thermal behavior of $(NH_4)_2MoO_2S_2$ and $(NH_4)_2WO_2S_2$ has been examined in vacuum thermogravimetrically[177]. Deflections in the TGA curve were obtained at $\approx$100, $\approx$300, and $\approx$1200 °C. The first step corresponds to the loss of ammonia; however, the nature of the products obtained from this decomposition were not examined. The thermal decomposition of tetraethylammonium tetrathiomolybdate was studied thermogravimetrically and found to proceed according to the equation[166]:

$$[(C_2H_5)_4N]_2MoS_4 \longrightarrow MoS_3 + (C_2H_5)_3N + (C_2H_5)_2S.$$

This decomposition of the tetrathio salt occurs at $\approx$180 °C.

The decomposition of $(NH_4)_2WS_4$ and the nature of the resulting products has been studied in detail by X-ray, DTA, opical microscopy, and X-ray diffraction[198,199]. The process occurring for the overall reaction is:

$$(NH_4)_2WS_4 \longrightarrow WS_3 + H_2S + 2\ NH_3$$

The decomposition depicted by the above-mentioned equation begins at $\approx$120 °C or higher and depends on the pressure and heating rate. The tungsten sulfide formed is poorly crystallized and has the composition $WS_{2.6-3.3}$. It is black and highly porous. At 330 °C, crystallization of WS_2 begins. The surface areas of WS_3 and WS_2 thus produced are 50–70 m^2/g[198].

Spectra

The infrared, Raman, visible and ultraviolet spectra of the thiomolybdate and thiotungstate species have been examined in detail and are given in several of the papers cited. The Mo-S and W-S stretching frequencies in these compounds are in the vicinity of 480–450 cm^{-1}. The infrared and Raman spectra of some thiochalcogenide anions reported by Müller[200] are given in Tables 10 and 11.

The infrared and Raman spectra of the ions $MoO_2S_2^{-2}$ and $WO_2S_2^{-2}$ in solids have also been given[201]. The Raman spectra of the ions $MoOS_3^{-2}$, WOS_3^{-2}, and MoS_4^{-2} in solids have been reported[202]. These results indicate that the symmetries of these anions are: $MoS_4^{-2}(T_d)$, $MoOS_3^{-2}(C_{3v})$, and $MoO_2S_2^{-2}(C_{2v})$.

The electronic spectra of the MoO_3S^{-2}, WO_3S^{-2}, and of the $MoO_2S_2^{-2}$ and $WO_2S_2^{-2}$ anions have been reported[171,177,203]. The electronic spectra of $MoO_2S_2^{-2}$, $MoOS_3^{-2}$, and MoS_4^{-2} have been examined in detail[204]. Those of the MoO_3Se^{-2} and WO_3Se^{-2} anions[173] and of the $MoOS_2Se^{-2}$ and $WO_2S_2Se^{-2}$ species have also been

Tab. 10. Vibrational frequencies of tetrachalcometallate anions with T_d symmetry (cm^{-1})

Anion	$\nu_1(A_1)$	$\nu_2(E)$	$\nu_3(F_2)$	$\nu_4(F_2)$	Ref.[2]
MoO_4^{-2} [1]	897	(318)	841	318	(1)
MoS_4^{-2}	460	(195)	480	195	(2)
$MoSe_4^{-2}$	255	(120)	340	120	
WO_4^{-2} [1]	931	(324)	833	324	(1)
WS_4^{-2}	485	(185)	465	185	(2)
WSe_4^{-2}	278	(115)	310	115	

[1]) Raman spectra in water solution.

[2]) (1) H. Siebert: Anwendungen der Schwingungsspektroskopie in der anorganischen Chemie, Springer 1966
(2) A. Müller, B. Krebs, W. Rittner, M. Stockburger: Ber. Bunsenges. Phys. Chem. *71*, 182 (1967).

Tab. 11. Raman Spectra of $(NH_4)_2MoO_2S_2$ and $(NH_4)_2WO_2S_2$ showing the range of Internal vibrations of the anions (frequencies in cm^{-1})

$MoO_2S_2^{-2}$	$WO_2S_2^{-2}$	CrO_2Cl_2 [1])	Frequencies for XY_2Z_2	
820 (m)	850 (m)	981	$\nu_1(A_1)$	$\nu_s(XY)$
473 (sst)	475 (sst)	465	$\nu_2(A_1)$	$\nu_s(XZ)$
310 (s)	310 (s)	356	$\nu_3(A_1)$	$\delta(XY_2)$
200 (st)	196 (st)	140	$\nu_4(A_1)$	$\delta(XZ_2)$
270 (s)	280 (s)	224	$\nu_5(A_2)$	τ
800 (m)	810 (m)	995	$\nu_6(B_1)$	$\nu_{as}(XY)$
248 (s)	235 (s)	211	$\nu_7(B_1)$	$\delta(XYZ)$
504 (m)	~475 (Sch)	496	$\nu_8(B_2)$	$\nu_{as}(XZ)$
270 (s)	280 (s)	257	$\nu_9(B_2)$	$\delta(YXZ)$

[1]) From H. Siebert: Anwendungen der Schwingungsspektroskopie in der anorganischen Chemie. Springer 1966.

reported[181)]. The reflectance spectra of the solids $(NH_4)_2MoS_4$ and $(NH_4)_2WS_4$ have been given[205)]. The electronic spectra of the deeply colored ions $MoOS_3^{-2}$ and WOS_3^{-2} in water solution were measured in the range 10,000 to 45,000 cm^{-1} and discussed[206)]. These spectra are due to charge transfer involving the S $\longrightarrow$ metal or Se $\longrightarrow$ metal charge transfer transitions.

Miscellaneous

Insoluble salts of the tetrathiomolybdate and tetrathiotungstate anions with coordination metal cations have been prepared and used for the gravimetric determination of molybdenum and tungsten[207,208]. For example, stable salts thus prepared are $[Cr(NH_3)_6Cl]MoS_4$ and $[Cr(NH_3)_6]_2(WS_4)_3$. Noncorrosive lubricants for metal surfaces were prepared by mixing 40 parts $PbMoS_4$ ($PbS + MoS_3$) and 60 parts MoS_2; the lead "thiomolybdate" was prepared from lead acetate and ammonium tetrathiomolybdate[209]. Other metal "thiomolybdates" used for similar purposes in addition to lead include thorium, cadmium, and zinc[210]. The reduction of thiomolybdate solutions at high temperatures using pressurized hydrogen and carbon monoxide has been studied. The products obtained are mixtures of a molybdenum hydrosulfide and hydroxide. The reduction with hydrogen was found to be autocatalytic[211]. Adherent lubricating MoS_2 films have been reported to form on metal surfaces by the decomposition of alkylammonium tetrathiomolybdates[212]. Synthetic lubricants containing MoS_2 were obtained by mixing heat-resistant high molecular weight polymers with MoS_2 of 1–5 particle size. The latter was prepared by reduction of $(NH_4)_2MoS_4$ at 600 °C in a hydrogen stream and the product ground to the desired size[213].

Hydrodesulfurization

The properties of the molybdenum-sulfur compounds described are of direct importance to hydrodesulfurization catalysis. Hydrodesulfurization and the nature of the catalysts used have been adequately described elsewhere and will not be presented here. References in which this system is described in detail include Refs.[161] and [214] through[221].

References

1) Kirk, R. E., Othmer, D. F. (eds.): Encyclopedia of chemical technology. 2nd edit., Vol. 13. New York: John Wiley and Sons, Inc. 1967, pp. 645–659

2) Killeffer, D. H., Linz, A.: Molybdenum compounds. New York: Interscience Publishers, 1952, Chapters 2, 4, and 5

3) Tsigdinos, G. A.: Heteropoly compounds of molybdenum and tungsten. Climax Molybdenum Company Bulletin Cdb-12 (Revised), November 1969; see also Refs. [226] and [227]

4) Tsigdinos, G. A., Hallada, C. J.: Isopoly compounds of molybdenum, tungsten, and vanadium. Climax Molybdenum Company Bulletin Cdb-14, February 1969; see also Ref. [228]

5) Properties of the simple molybdates. Climax Molybdenum Company Bulletin Cdb-4, October 1962

6) Mellor, J. W.: A comprehensive treatise on inorganic and theoretical chemistry. Vol. XI. London: Longmans, Green, and Company, 1931, pp. 640–658

7) Hallada, C. J.: Climax molybdenum company, unpublished results

8) Espelund, A. W.: Acta Chem. Scan. *21*, 839 (1967)

9) Takeuchi, Y., Nowacki, W.: Schweiz. Mineral. Petrogr. Mitt. *44*, 105 (1964)

10) Wickman, F. E., Smith, D. K.: Am. Mineral. *55*, 1843 (1970)

11) Frondel, J. W., Wickman, F. E.: Am. Mineral. *55*, 1857 (1970)

12) Properties of molybdenum disulfide. Climax Molybdenum Company Bulletin Cdb-5a, February 1962
13) Romanowskii, W.: Rocz. Chem. *37,* 1077 (1963)
14) Seshadri, K. S., Massoth, F. E., Petrakis, L.: J. Catal. *19,* 95 (1970)
15) Wildervanck, J. C., Jellinek, F.: Z. Anorg. Allgem. Chem. *328,* 309 (1964)
16) Stubbles, J. R., Richardson, F. D.: Trans. Faraday Soc. *56,* 1460 (1960)
17) Hilli, A. A., Evans, B. L.: J. Cryst. Growth. *15,* 93 (1972)
18) Mering, J., Levialdi, A., C. R. Acad. Sci., Paris *213,* 798 (1941)
19) Rode, E. Y., Lebedev, B. A.: Russ. J. Inorg. Chem. *6,* 608 (1961)
20) Ratnasamy, P., Leonard, A. J.: Catal. Rev. *6,* 293 (1972)
21) Ratnasamy, P., Rodrique, L., Leonard, A. J.: J. Phys. Chem. *77,* 2242 (1973)
22) Bell, R. E., Herfert, R. E.: J. Am. Chem. Soc. *79,* 3351 (1957)
23) Silverman, M.: Inorg. Chem. *6,* 1063 (1967)
24) Arutyunyan, L. A., Khurshudyan, E. K.: Geokhimiya *1966,* 650
25) Mishin, I. V., Feodot'ev, K. M.: Ocherki, Geokhim. Rtuti, Molibdena Sery Gidroterm. Protsesse *1970,* 102
26) Zelikman, A. N., Krein, O. E.: Christye Metal. i Poluprovodn., Trudy 1-oi [Pervoi] Mezhvuz. Konf., Moscow, 1957, 336-43 (Pub. 1959); CA *55,* 7771c (1961)
27) Zelikman, A. N., Krein, O. E., Teslitskaya, M. V.: Khal'Kogenidy (*2*), 47 (1970)
28) Andrieux, J. L., Weiss, G.: Bull. Soc. Chim. Fr. *1948,* 598
29) Weiss, G.: Ann. Chim. (Paris) *1,* 446 (1966)
30) Zelikman, A. N., Christyakov, Y. D., Indenbaum, G. V., Krein, E. O.: Kristallografiya *6,* 389 (1961)
31) Domsa, A., Szabo, L, Spirchez, Z.: Metalurgia (Bucharest) *16,* 162 (1964)
32) Malinov, D.: Mashinostroene *17,* 445 (1968)
33) LeBoete, F., Mathiron, C., Toesca, S., Delafosse, D., Colson, J. C.: Bull. Soc. Chim. Fr. *1969,* 3869
34) Von Hahn, F. V.: Kolloid-Z., Special No., April 1 (1925), 277-86; CA *19,* 3048^5 (1925)
35) Dickinson, R. G., Pauling, L.: J. Am. Chem. Soc. *45,* 1466 (1923)
36) Jellinek, F., Brauer, G., Müller, H.: Nature *185,* 376 (1960)
37) Traill, R. V.: Can. Mineralogist 7, 524 (1963)
38) Graeser, S.: Schweiz. Mineral. Petrogr. Mitt. *44,* 121 (1964)
39) Clark, A. H.: Mineralogy Mag. *35,* 69 (1965)
40) Kruglova, V. G., Sidorenko, G. A., Populpanova, L. I.: Tr. Mineralog. Muzeya, Akad. Nauk. SSSR No. 16, 233 (1965)
41) Sidorenko, G. A.: Rentgenogr. Mineral'n. Syr'ya, Vses. Nauchn. Issled. Inst. Mineral'n. Syr'ya Akad. Nauk SSSR, No. *5,* 33 (1966); CA *65,* 161721 (1966)
42) Semiletov, S. A.: Kristallographia *6,* 536 (1961)
43) Tang, Au-Chin, Liu, Jo-Chuang: J. Chinese Chem. Soc. *18,* 53 (1951)
44) Barinskii, R. L.: Dokl. Akad. Nauk SSSR *83,* 381 (1952)
45) Barinskii, R. L.: Vainshtein, E. E.: Izv. Akad. Nauk SSSR, Ser. Fiz. *21,* 1387 (1957)
46) Cannon, P.: Nature *183,* 1612 (1959)
47) Kalikhman, V. L., Radzikovskaya, S. V., Bukhanevich, V. F., Gladchenko, E. P.: Porosh. Met. *10,* 55 (1970)
48) Dahl, J. M.: Climax molybdenum company, unpublished results
49) Plaskin, I. N., Shafeev, R. S.: Dokl. Akad. Nauk SSSR *132,* 399 (1960)
50) Oganesyan, V. K., Rud, B. M.: Porosh. Met., Akad. Nauk Ukr. SSR *5,* 54 (1965)
51) Zelikman, A. N., Varonov, B. K., Dudkin, L. D.: Izv. Akad. Nauk SSR, Neorgan. Mater. *7,* 428, 433 (1971)
52) Spackman, J. W. C.: Nature *198,* 1266 (1963)
53) Frindt, R. F., Yoffe, A. D.: Proc. Roy. Soc. (London) Ser. A, *273,* 69 (1963)
54) Frindt, R. F.: Phys. Rev. *140* (2A), 536 (1965)
55) Gillet, M.: Bull. Microscop. Appl. *12,* 121 (1962)
56) Evans, B. L., Thompson, K. T.: Brit. J. Appl. Phys. *1,* 1619 (1968)
57) Zalevskii, B. K., Opalovskii, A. A., Sobolev, V. V., Fedorov, V. E.: Izv. Sib. Otd. Akad. Nauk SSSR, Ser. Khim. Nauk *1967,* 14

58) Clark, A., Williams, R. H.: Brit. J. Appl. Phys. *1,* 1222 (1968)
59) Dudnik, E. M., Oganesyan, V. K.: Porosh. Met. *6* (2), 60 (1966)
60) Molysulfide Newsletter: Climax molybdenum company. Vol. 12, No. 1, April 1969
61) Smith, D. F., Brown, D., Dworkin, A. S., Sasmor, D. J., Van Artsdalen, E. R.: J. Am. Chem. Soc. *78,* 1533 (1956)
62) Westrum, E. F., Jr., McBride, J. J.: Phys. Rev. *98,* 270 (1954)
63) Kelley, K. K.: Thermodynamic properties of molybdenum compounds. Climax Molybdenum Company Bulletin Cdb-2, 1954
64) Parravano, N., Malquori, G.: Atti accad. Lincei [6] 7, 109 (1928)
65) McCabe, C. L.: J. Metals *7,* 61 (1961)
66) Scholz, W. G., Doane, D. V., Timmons, G. A.: Trans. Met. Soc. AIME *221,* 356 (1961)
67) Schaefer, S. C., Larson, A. H., Schlechten, A. W.: Trans. Met. Soc. AIME *230,* 594 (1964)
68) Gorokh, A. V., Klokotina, L. I., Rispel, K. N.: Dokl. Akad. Nauk SSSR *158,* 1183 (1964)
69) Gorokh, A. V., Klokotina, L. I., Rispel, K. N., Demidov, Y. Y.: Teor. Prakt. Met. No. *8,* 163 (1966)
70) Opalovskii, A. A., Fedorov, V. E.: Dokl. Akad. Nauk, SSSR *163,* 900 (1965)
71) Amman, P. R., Loose, T. A.: Met. Trans. *2,* 889 (1971)
72) Zelikman, A. N., Belyaevskaya, V.: Zh. Neorg. Khim. *1,* 2245 (1956)
73) Zelikman, A. N., Tumin, N. A, Bryukuin, V. A.: Izu. Vyssh. Vcheb, Zaved., Tsued. Met. *14,* 59 (1971)
74) JANAF thermochemical tables and addenda. Midland, Michigan: Dow Chemical Company, 1965
75) Godfrey, D., Nelson, E. C.: NASA Technical Note No. 1882 (1949)
76) Lavik, M. T., Medved, T. M., More, G. D.: ASLE Trans. *11,* 44 (1968)
77) Vasilev, K., Kuzmanov, B., Dimitrov, R.: Khim. Ind. (Sofia) (*5*), 203 (1968)
78) Fedulov, O. V., Taranenko, B. I., Ponomarey, V. D., Svechkova, L. Y.: *Sb. Slatei Aspin. Soiskatelei, Min. Vyssh. Sredn. Spets. Obraz. Kaz. SSR, Met. Obogashch* (*2*), 186 (1966)
79) Vikulov, A. I., Nesmeyanova, G. M.: Zh. Prikl. Khim. (Leningrad) *43,* 965 (1970)
80) Undasynova, Z. D., Ponomareva, E. I.: Izv. Akad. Nauk Kaz. SSR, Ser. Teckhn. i Khim. Nauk (*2*), 53 (1963)
81) Ponomareva, E. I., Baikenov, K. I.: Vestn. Akad. Nauk Kaz. SSR *21,* 42 (1965)
82) Dresher, W. H., Wadsworth, M. E., Fassell, W. M., Jr.: Trans. AIME *206*, 794 (1956)
83) Pitts, J. J., Jache, A. W.: Inorg. Chem. *7,* 1661 (1968)
84) Sinakevich, A. S.: Sbornik Nauch. Trudov Irkutsk. Inst. Redkikh Metal (*7*), 171 (1958)
85) Rza-Zade, P. F., Samedov, M. A.: Izv. Akad. Nauk Azerbaidzhan SSR *1956,* 49
86) Krey, C., Poole, H. G., Shelton, S. M.: NASA Doc. N63-18058, 1963, 98 pp.
87) Obolonchik, V. A., Prokoshina, L. M., Radzikovskaya, S. V.: Porosh. Met. *9,* 53 (1969)
88) Deev, V. I., Smirnov, V. I.: Tsvetn. Metal *37,* 63 (1964)
89) Lebedev, K. B., Tyurekhodzhaeva, T. S.: Tr. Inst. Met. i Obogashch., Akad. Nauk Kaz. SSR *4,* 170 (1962)
90) Polyvyannyi, I. R., Milyutina, N. A.: Tr. Inst. Met. i Obogashch., Akad. Nauk Kaz. SSR *21,* 3 (1967)
91) Yagn, N. I., Sokolova, B. G.: Zapiski Vsesoyuz. Mineral. Obshchestva *88,* 72 (1959)
92) Cannon, P.: J. Chim. Phys. *58,* 126 (1961)
93) Rüdorff, W.: Angew. Chem. *71,* 487 (1959)
94) Rüdorff, W.: Chimia *19,* 489 (1965)
95) Rüdorff, W., Ostertag, W.: Proc. Conf. Rare Earth Res., 4th, Phoenix, Arizona, 1965, pp. 117-24
96) Farragher, A. L., Cossee, P.: Catal., Proc. Int. Congr., 5th, *2,* 1301 (1972) (Pub. 1973)
97) Feng, I-Ming: Lubrication Eng. *8,* 285 (1952)
98) McCabe, J. T.: Molybdenum disulfide – Its Role in Lubrication. Presented at the Symposium of the International Industrial Lubrication Exhibition, Westminster, London: Royal Horticultural Society's New Hall, S. W. I., March 8–11, 1965
99) Winer, W. O.: Wear *10,* 422 (1967)
100) Braithwaite, E. R.: Solid lubricants and surfaces. New York: The MacMillan Company 1964

101) Claus, F. J.: Solid lubricants and self-lubricating solids. New York: Academic Press, Inc. 1972
102) ASLE Proceedings – International Conference on Solid Lubrication, 1971, Denver, Colorado, August 24–27, 1971
103) Molybdenum catalyst bibliography (1950–1964). Climax Molybdenum Company, Compiled by Warner, P. O., Barry, H. F.
104) Molybdenum catalyst bibliography (1964–1967). Supplement 1, Climax Molybdenum Company, Compiled by Losey, E. N., Means, D. K.
105) Molybdenum catalyst bibliography (1967–1969). Supplement 2, Climax Molybdenum Company, Compiled by Rudolph, R.
106) Molybdenum catalyst bibliography (1970–1972). Supplement 3, Climax Molybdenum Company, Compiled by Tsigdinos, G. A.
107) Weisser, O., Landa, S.: Sulfide catalysts, their Properties and applications. New York: Pergamon Press 1973
108) Eggertsen, F. T., Roberts, R. M.: J. Phys. Chem. *63,* 1981 (1959)
109) Guichard, M.: Bull. Soc. Chim. Fr. *51,* 563 (1932)
110) Jellinek, F.: Nature *192,* 1065 (1961)
111) de Jonge, R., Pompa, T. J. A., Wiegers, G. A., Jellinek, F.: J. Solid State Chem. *2,* 88 (1970); Jellinek, F.: private communication 1971
112) Gorokh, A. V., Rusakov, L. N., Savinskaya, A. A.: Dokl. Akad. Nauk SSSR *156,* 541 (1964)
113) Hager, J. P., Elliott, J. E.: Trans. Met. Soc. AIME *239,* 513 (1967)
114) Wilcox, D. E., Bromley, L. A.: Ind. Eng. Chem. *55,* 32 (1963)
115) German Patent 1,099,433 (to Naxos-Union Schleifmittel- und Schleifmaschinenfabrik), February 9, 1961; CA *56,* 8324e (1962)
116) Taimni, I. K., Agarwal, R. P.: Anal. Chim. Acta *9,* 203 (1953)
117) Taimni, I. K., Tandon, S. N.: Anal. Chim. Acta *22,* 34 (1960)
118) Haran, M., Srivastana, N., Ghosh, S.: Proc. Natl. Acad. Sci. India *A29,* Pt2, 178 (1960)
119) Zvorykin, A. Y., Perel'man, F. M., Tarasov, V. V.: Zh. Neorg. Khim. *6,* 1994 (1961)
120) DeBucquet, L., Velluz, L.: Bull. Soc. Chim. Fr. *51,* 1571 (1932)
121) Biltz, W., Kocher, A.: A. Anorg. Allgem. Chem. *248,* 172 (1941)
122) Mawrow, F., Nokolow, M.: Z. Anorg. Chem. *95,* 188 (1916)
123) Berzelius, J. J.: Ann. Chim. Phys. *29,* 369 (1825)
124) Towle, L. L., Oberbeek, V., Brown, B. E., Stajdohar, R. E.: Science *154,* 895 (1966)
125) Opalovskii, A. A., Fedorov, V. E.: Dokl. Akad. Nauk SSR *163,* 1163 (1965)
126) Wells, A. F.: Structural inorganic chemistry. Third edit. Oxford: The Clarendon Press 1962, pp. 457, 510
127) Bisson, E. E., Johnson, R. L., Swikert, M. A., Godfrey, D.: NACA Report *1956,* 1254 Washington, D.C.
128) Miklailov, A. S.: Geokhimiya *1962,* 818
129) Ong, J. N., Wadsworth, M. E., Farrell, W. M., Jr.: AIME *206,* 257 (1956)
130) Spengler, G., Weber, A.: Chem. Ber. *92,* 2163 (1959)
131) Badger, E. H. M., Griffith, R. H., Newling, W. B. S.: Proc. Roy. Soc. *A197,* 184 (1949)
132) Richardson, J. T.: Ind. Eng. Chem. Fundam. *3,* 154 (1964)
133) ter Meulen, H.: Chem. Weekbl. *22,* 218 (1925)
134) Cannon, P., Norton, N. F.: Nature *203,* 750 (1964)
135) Smith, F. F., Oberholtzer, V.: Z. Anorg. Chem. *5,* 66 (1894)
136) Glukhov, I. A.: Izvest. Otdel. Estestven. Nauk Akad. Nauk Tadzhik SSR (*24*), 21 (1957)
137) Fortunatov, N. S., Timoshchenko, N. I.: Ukr. Khim. Zh. *31,* 1078 (1965)
138) Funk, H., Berndt, K. H., Henze, G.: Wiss. Z. Univ. Halle-Wittenberg *6,* 815 (1957)
139) Rannou, J. P., Sergent, M.: C. R. Acad. Sci., Paris, Ser. C, *265,* 734 (1967)
140) Perrin-Billot, C., Perrin, A., Prigent, J.: Chem. Commun. *1970,* 679
141) Sharma, K. M., Anand, S. K., Multani, R. K., Jain, B. D.: Chem. Ind., (London) *1969,* 556
142) Britnell, D., Fowles, G. W. A., Mandyczewsky, R.: Chem. Commun. *1970,* 608
143) Drew, M. G. B., Mandyczewsky, R.: J. Chem. Soc. (*A*), 2815 (1970)
144) Opalovskii, A. A., Fedorov, V. E.: Dokl. Akad. Nauk SSSR *182,* 1095 (1968)

145) Sergent, M., Prigent, J.: C. R. Acad. Sci., Paris, Ser. C, *261*, 5135 (1965)
146) Bronger, W., Huster, J.: Naturwissenschaften *56*, 88 (1969)
147) Rudnev, N. A., Malofeyeva, G. I.: Talanta *11*, 531 (1964)
148) Chevrel, R., Sergent, M., Prigent, J.: C. R. Acad. Sci., Paris, Ser. C, *267*, 1135 (1968)
149) van den Berg, J. M.: Inorg. Chim. Acta *2*, 216 (1968)
150) Anzenhoffer, K., de Boer, J. J.: Acta Crystallogr. B *25*, 1419 (1969)
151) Guillevic, J., Chevrel, R., Sergent, M.: Bull. Soc. Mineral. Cristallogr. *93*, 495 (1970)
152) Guillevic, P. J., Le Marouille, J.-Y., Granjean, D.: Acta Crystallogr. B *30*, 111 (1974)
153) Chevrel, R., Guillevic, J., Sergent, M.: C. R. Acad. Sci., Paris, Ser. C, *271*, 1240 (1970)
154) Alderson, W. L., Maynard, J. T.: US 2,770,527 (to E. I. du Pont de Nemours and Co.), November 13, 1956
155) Chevrel, R., Sergent, M., Prigent, J.: J. Solid State Chem. *3*, 515 (1971)
156) Grover, B.: Neues Jahrb. Mineral. Monatsch. (7), 219 (1965)
157) Kellerund, G.: Carnegie Inst. Wash. Yr. Bk. *65*, 337 (1965–66)
158) LeBoete, F., Mathiron, C., Toesca, S., Delafosse, D., Colson, J.: Bull. Soc. Chim. Fr. *1969*, 3869
159) Engelhard, P. A., Trambouze, Y.: Bull. Soc. Chim. Fr. *1959*, 195
160) Engelhard, P.: Chim. Mod. *4*, 61 (1959)
161) Armour, A. W., Ashley, J. H., Mitchell, P. C. H.: Cobalt-Molybdenum-Alumina Hydrodesulfurization Catalysts. Presented before the Division of Petroleum Chemistry, Inc., ACS Meeting, Los Angeles, Calif., March 28–April 2, 1971
162) Berzelius, J. J.: Poggend. Ann. *7*, 270 (1826)
163) Krüss, G.: Liebigs Ann. Chem. *225*, 6 (1884)
164) Corleis, E.: Liebigs Ann. Chem. *232*, 254 (1886)
165) Müller, A., Diemann, E.: Z. Naturforsch. *23b*, 1607 (1968)
166) Leroy, M. J. F., Kaufmann, G., Charlionet, R., Rohmer, R.: C. R. Acad. Sci., Paris *263*, 601 (1966)
167) Perel'man, F. M., Zvorykin, A. Y., Tarasov, V. V., Demina, G. A.: Russ. J. Inorg. Chem. *6*, 1023 (1961)
168) Ponomarev, V. D., Buketov, E. A.: Sbornik Nauch. Trudov, Kazakh. Gorno-Met. Inst., Geol., Gornoe Delo, Met. (*16*), 369 (1959); CA *55*, 9135a (1959)
169) Bernard, J. C., Tridot, G.: Bull. Soc. Chim. Fr. (*1961*), 810
170) Bernard, J. C., Tridot, G.: Bull. Soc. Chim. Fr. (*1961*), 819
171) Aymomino, P. J., Ranade, A. C., Müller, A.: Z. Anorg. Allgem. Chem. *371*, 295 (1969)
172) Petit, G., Bourlange, C., Seyyedi, A.: C. R. Acad. Sci., Paris *255*, 2238 (1962)
173) Müller, A., Ranade, A. C., Rittner, W.: Z. Anorg. Allgem. Chem. *380*, 76 (1971)
174) Spivac, B., Dori, Z.: J. Chem. Soc., Chem. Commun. (*23*), 909 (1973)
175) Müller, A., Krebs, B., Diemann, E.: Z. Anorg. Allgem. Chem. *353*, 259 (1967)
176) Müller, A., Krebs, B.: Z. Anorg. Allgem. Chem. *342*, 182 (1966)
177) Müller, A., Diemann, E., Baran, E. J.: Z. Anorg. Allgem. Chem. *375*, 87 (1970)
178) Tsigdinos, G. A.: Climax Molybdenum Company, unpublished results
179) Müller, A., Diemann, E., Krebs, B., Leroy, M. J. F.: Angew. Chem. Internat. Ed. *7*, 817 (1968)
180) Müller, A., Diemann, E., Heidborn, U.: Z. Anorg. Allgem. Chem. *371*, 136 (1969)
181) Müller, A., Diemann, E.: Chem. Ber. *102*, 2603 (1969)
182) Müller, A., Diemann, E., Schulze, H.: Z. Anorg. Allgem. Chem. *376*, 120 (1970)
183) Müller, A., Diemann, E., Heidborn, U.: Z. Anorg. Allgem. Chem. *376*, 125 (1970)
184) Müller, A., Sievert, W., Schulze, H.: Z. Naturforsch. *B27*, 720 (1972)
185) Clark, G. M., Doyle, W. P.: J. Inorg. Nucl. Chem. *28*, 381 (1966)
186) Saxena, R. S., Jain, M. C., Mittal, M. C.: Monatsh. Chem. *99*, 530 (1968)
187) Ziegler, M., Glemser, O.: Angew. Chem. *68*, 620 (1956)
188) Gattow, G., Franke, A.: Z. Anorg. Allgem. Chem. *352*, 11 (1967)
189) Saxena, R. S., Jain, M. C., Mittal, M. L.: Aust. J. Chem. *21*, 91 (1968)
190) Aveston, J., Anacker, E. W., Johnson, S. J.: Inorg. Chem. *3*, 735 (1964)
191) Gattow, G.: Naturwissenschaften *46*, 425 (1959)
192) Savari, K.: Acta Crystallogr. *16*, 719 (1963)

193) Schäfer, H., Schäfer, G., Weiss, A.: Z. Naturforsch. *19b*, 76 (1964)
194) Binnie, W. P., Ridman, M. J., Mallo, W. J.: Inorg. Chem. *9*, 1449 (1970)
195) Koz'min, P. A., Popova, Z. V.: Zh. Strukt. Khim. *12*, 99 (1971)
196) Müller, A., Diemann, E.: Chem. Commun. *1971*, 65
197) Müller, A., Diemann, E., Heidborn, U.: Z. Naturforsch. *24b*, 1482 (1969)
198) Voorhoeve, R. J., Wolters, H. B. M.: Z. Anorg. Allgem. Chem. *376*, 165 (1970)
199) Voorhoeve, R. J., Wolters, H. B. M.: Z. Anorg. Allgem. Chem. *380*, 326 (1971)
200) Müller, A., Krebs, B., Kebabcioylu, R., Stockburger, M., Glemser, O.: *Spectrochim. Acta 24A*, 1831 (1968)
201) Leroy, M. J. F., Kaufmann, G.: Bull. Soc. Chim. Fr. *1968*, 3586
202) Leroy, M. J. F., Kaufmann, G.: Bull. Soc. Chim. Fr. *1968*, 4028
203) Müller, A., Krebs., B., Rittner, W., Stockburger, M.: Ber. Bunsengesellschaft physik. Chem. *71*, 182 (1967)
204) Bartecki, A., Dembicka, D.: Inorg. Chim. Acta *1*, 610 (1973)
205) Companion, A. L., Mackin, M.: J. Chim. Phys. *42*, 4219 (1965)
206) Diemann, E., Müller, A.: Spectrochim. Acta *26A*, 215 (1970)
207) Spacu, G., Spacu, P., Gheorghiu, C.: Acad. rep. populare Romine, Studii cercetari chim. *5*, 169 (1957); CA *51*, 17549n (1957)
208) Spacu, G., Gheorghiu, C.: Commun. acad. rep. populare Romine *5*, 853 (1955); CA *50*, 16547b (1956)
209) Lerer, M.: Fr. 1,423,641 (to Institut Français du Petrole, des Carburants et Lubrifiants), January 7, 1966; CA *65*, 8623f (1966)
210) Hugel, G.: Compt. rend. congr. intern. chim. ind., 31st, Liège, 1958 (Pub. as Ind. chim. Belge, Suppl.) *1*, 348 (1959); CA *54*, 9265b (1960)
211) Warren, I. H., Okita, Y.: I and EC Fundamentals *12*, 342 (1973)
212) Spengler, G., Hohn, H.: US 2,905,574 (to Alpha Molykote Corp.) September 22, 1959
213) Wolf, W. Jantsch, F.: Ger. 1,003,385 (to Badische Anilin and Soda-Fabrik Akt.-Ges.) February 28, 1957; CA *53*, p22896e (1959)
214) McKinley, J. B.: The hydrodesulfurization of liquid petroleum fractions. Emmett, P. H. (ed.): Catalysis, Vol. V. New York: Reinhold Publishing Corporation 1957, pp. 405–526
215) Schuman, S. C., Shobit, H.: Hydrodesulfurization. Catal. Rev. *4*, 245–317 (1970)
216) Inoguchi, M.: Catalysts for the direct hydrodesulfurization of residual oils. Nenryo Kyokaishi *48*, 792–805 (1969)
217) Mitchell, P. C. H.: The chemistry of some hydrodesulfurization catalysts containing molybdenum. Climax Molybdenum Company Literature C-29, 1967
218) Lipsch, J. M. J. G.: The $CoO-MoO_3-Al_2O_3$ catalyst. Doctoral Thesis, Technological University, Eindhoven, The Netherlands 1969. Climax Literature C-37
219) Lipsch, J. M. J. G., Schuit, G. C. A.: J. Catal. *15*, 163, 174, 179 (1969)
220) Ashley, J. H., Mitchell, P. C. H.: J. Chem. Soc. *(A)*, 2821 (1968)
221) Ashley, J. H., Mitchell, P. C. H.: J. Chem. Soc. *(A)*, 2730 (1969)
222) Chevrel, R., Sergent, M., Prigent, J.: Mater. Res. Bull. *9*, 1487 (1974)
223) Whittingham, M. S.: J. Electrochem. Soc. *123*, 315 (1976)
224) Whittingham, M. S.: Science *192*, 1126 (1976)
225) Whittingham, M. S., Gamble, F. R.: Mat. Res. Bull. *10*, 363 (1975)
226) Tsigdinos, G. A.: Heteropoly-Verbindungen. Methodicum Chimicum, Niedenzu, K., Zimmer, H.: (ed.). Vol. 8, Chapter 32. Stuttgart: Georg Thieme Verlag 1974
227) Tsigdinos, G. A.: Heteropoly compounds, methodicum chimicum (English Edition). Niedenzu, K., Zimmer, H., (ed). Vol. 8, Chapter 32. New York: Academic Press 1976
228) Tytko, K.-H., Glemser, O.: Isopolymolybdates and isopolytungstates. Emeleus, J. H., Sharpe, A. G. (ed.). Advances in inorganic and radiochemistry, Vol. 19. Academic Press: New York, 1976

Received February 15, 1978

High-Temperature Metal Sulfide Chemistry

Günter H. Moh

Mineralogisch-Petrographisches Institut der Universität Heidelberg, Im Neuenheimer Feld 236, 6900 Heidelberg 1, Germany

Table of Contents

Introduction

Within this century, great advances have been made in investigating phase stability and phase relations of various multicomponent systems. Petrologists and ceramists have examined numerous silicate and oxide systems of chemical, geological, and economic interest and determined the compositions of minerals and high temperature-resistant refractory oxides of technical importance. Metallurgists and extraction mineralogists have studied intermetallic systems including those which produce intermetallic alloys with the elements nitrogen, carbon, or sulfur. New methods and tools have been developed for experimentation within larger temperature and pressure ranges, incorporating analytical techniques such as improved reflected-light microscopy, X-ray diffraction/fluorescence spectrometry, electron probe microanalysis, and Mössbauer spectrometry.

Evacuated rigid silica glass tubes and welded collapsible gold tubes as reaction vessels have been used in the last few decades with spectacular results for investigating certain sulfide-type systems. The upper temperature range for the silica tube technique is approximately 1000–1200 °C. Owing to this limitation some important systems such as molybdenum-sulfur or tungsten-sulfur, and mutual reactions between high temperature-resistant refractory sulfides could not be investigated systematically.

The work of laboratories of moderate resources is based upon resistance furnace methods. More elaborate methods, such as high-frequency induction, cooled-crucible arc, electron bombardment, or mirror furnaces, with their resulting higher temperature ranges are financially beyond the reach of these laboratories.

Recently a fairly inexpensive way of high-temperature experimentation has been found to investigate refractory sulfides and related multicomponent systems up to temperatures of nearly 2000 °C using resistance furnaces. These techniques are discussed below and applied to some sulfide systems, in particular of those metals which belong to the VI-B group. The binary systems chromium-sulfur, molybdenum-sulfur, tungsten-sulfur, as well as some other ternary and quaternary systems and their reactions are reviewed and completed within the limits of the new experimental procedure.

All the systems herein are represented as condensed systems (vapor is a stable coexisting phase, but its pressure is not constant). The term "vapor" is used in the sense defined by Morey[1)] who distinguished (p. 230) 1) liquid, 2) gas, 3) vapor, and 4) "fluid phase": "a vapor is a special case of a gas . . . which can be compressed to a liquid".

Experimental Procedures

Starting Materials

High-purity elements were used to synthesize the required binary and ternary compounds for further experimentation: chromium pellets of 99.999 + % from United Minerals & Chemical Corp.; molybdenum powder or rods of 99.99 + % and 99.95 + %

tungsten powder from Koch-Light Laboratories Ltd.; sulfur of 99.999 + % purity from ASARCO. High-purity copper, iron, and zinc were also used.

The chromium pellets were crushed into a fine powder in a steel mortar. Contamination by rubbed-off traces of magnetic metal powder from the mortar was eliminated by passing a strong hand magnet several times over the crushed product so that all steel traces were extracted. A similar procedure was used to prevent contamination of the molybdenum rods or other filed metals. Chromium, molybdenum, and tungsten were heated before use in a high vacuum (10^{-5} or better) at approximately 1000 °C for about 1 hr in order to expel intergranular hydrocarbon gases. Before this procedure the degassing apparatus was washed with pure hydrogen. After vacuum cooling, the materials were checked for possible surface oxidation with a magnifying binocular, and only the products without visible tarnish were subsequently used.

Techniques

The experimental results discussed below are basically derived from laboratory techniques for dry sulfide research as described in great detail by Kullerud[2]. At least these syntheses of various compounds were used for further high-temperature reactions. For the high-temperature procedures, the normal techniques with evacuated rigid silica glass tubes were modified to make them suitable for experimentation above 1200 °C. It was found that thick-walled high-purity silica glass tubes can be used successfully for short periods of reaction time up to 1500 or 1600 °C. They will even stand relatively high vapor pressures. Breakage at these high temperatures is due to devitrification. With increasing temperatures and reaction times, the vessels will devitrify, particularly in those runs that contain molybdenum, tungsten, etc. which exaggerate this crystallization process. To avoid this disadvantage, an inner shielding is used to protect the actual tube vessel, as illustrated in Fig. 1.

A flame-polished open-ended silica glass tube holds the charge. A silica glass rod is placed on top of the charge, and this tube is inserted into a larger but closely fitting normal tube (see Fig. 1, A–B). This assembly is evacuated and sealed (Fig. 1C) in the way described by Kullerud[2]. Subsequently that tube-in-tube arrangement is placed in a third closely fitting but thick-walled tube open at one end (Fig. 1D). Special care is necessary to avoid contamination.

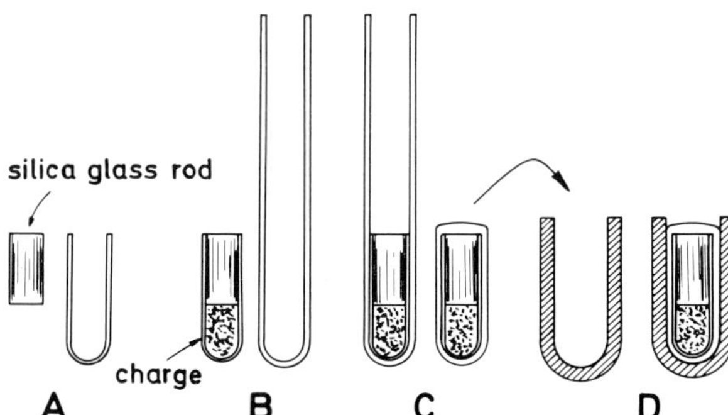

Fig. 1. Illustration of "tube-in-tube-in-tube" reaction vessels applicable for high-temperature sulfide experimentation. The inner reaction vessel (A) is 2 to 4 mm I.D. (wall thickness 1 mm) with a total length not exceeding 25 to 30 mm. The dimensions of the closely fitting second tube and the outside vessel are consequently larger

All parts including the welded seal must be very clean. Another important feature is the necessity of a high vacuum in the sealed tube. The gas pressure (excluding the partial sulfur vapor pressure of the sulfidic charge) in a vessel with a poor vacuum may exceed one atmosphere far below the desired reaction temperature. The tube-in-tube-in-tube assembly is placed vertically in a graphite block crucible (cf. Fig. 2C). Upon heating the increased internal pressure may expand the tube-in-tube assembly against the thick-walled block which holds the pressure. Carefully prepared assemblies can be used for 1 hr or more at 1400–1500 °C. Successful experiments were performed at 1600 °C for shorter periods. Normally a period of minutes is sufficient to obtain equilibrium with sulfidic charges at high temperatures. During experimentation, only the inner and partly the outer of the three tubes will devitrify, owing to contamination from the charge or from the graphite crucible, respectively. A Tammann furnace was employed for these experiments.

To perform quenching experiments utilizing a horizontal furnace, the tube-in-tube assembly is modified slightly, as follows: the normally evacuated and sealed inner tube-in-tube is inserted into an open-ended closely fitting thick-walled tube. This assembly is placed into another closely fitting tube, which is evacuated and sealed. These runs were successfully carried out for experiments in platinum or in platinum-rhodium wound horizontal furnaces with temperatures beyond 1400 °C. After a predetermined heating period the probes were quenched in either air or water.

For experiments with temperatures above the melting point of silica glass the entire tube-in-tube run was inserted into a closely fitting graphite crucible, as illustrated in Fig. 2 A–E. At elevated temperatures, the sulfidic charge is protected by a homogeneous silica liquid (see Fig. 2E). Owing to the low vapor pressure of the investigated sulfides and the high viscosity

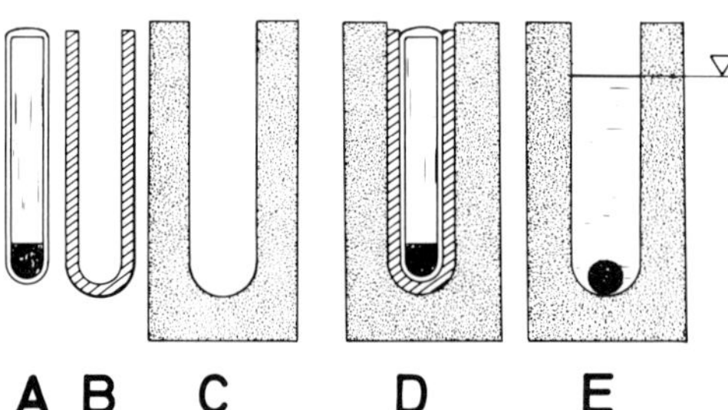

Fig. 2. High-temperature sulfide experimentation vessel as described in the text. The inner tube (A) normally does not exceed 3 mm I.D. (1 mm wall thickness) with a length of approx. 50 mm. The tube is surrounded by the silica glass vessel and graphite crucible

of the silica liquid, the charge remains near the bottom of the crucible, but completely surrounded by a gas bubble of its own vapor, thus preventing the surrounding silica liquid from "moistening" the charge. The sulfide vapor pressure can be balanced by the outside gas pressure. In our experiments, the vapor pressure of the charge was limited to the atmospheric pressure by washing the furnace continuously with completely oxygen-free dry nitrogen. The experiments were carried out with temperatures up to approximately 1900 °C. At higher temperatures SiO_2 (melt) is reduced by the graphite crucible to SiO.

The heating time in the Tammann furnace up to 1800 or 1900 °C was 2 to 3 h, and the furnace cools to room temperature in about 1/2 h. After switching off the electric current at reaction temperature, the sample is "quenched" by a very sudden cooling of the first few hundreds of degrees, at which time the silica liquid chills. The graphite crucible was crushed after each run, and the silica glass regulus including the charge was easily separated for the following room temperature examinations: macroscopic, reflected-light microscopy, and X-ray diffraction. Polished sections for the microscopic studies were prepared as described by Kullerud[2)]. In none of the experiments with the Tammann furnace was the formation of carbides observed. The Tammann furnace temperatures were measured *via* optical pyrometry with an accuracy of ±50 °C. The Tammann furnace used for the above experimental procedures is situated in the Anorganisch-Chemisches Institut der Universität Heidelberg. The platinum wound furnaces were made by ourselves and are similar to the type of nichrome wound furnaces described by Kullerud[2)]. The temperatures were measured by means of Pt/Pt-Rh thermocouples, which have an accuracy better than ±5 °C.

In many cases, high-temperature modifications of sulfidic compounds cannot be quenched for room temperature examination. Inversion twinnings, crystal morphology, or other crystallographic features may indicate the appearance of polymorphism. Under these circumstances differential thermal analysis (DTA) can be suitable for the determination of the exact phase transition temperatures. DTA determinations are practically valuable if used in conjunction with high-temperature X-ray diffraction methods. DTA apparatus can operate up to 1100 °C and can be specially designed for sulfides[2–4]: individual experimental techniques are included in these references.

Recently it became possible to carry out DTA experiments with high temperature-resistant sulfides beyond 1100 °C. An apparatus from the Soviet Union intended for metallurgic research up to 2350 °C could be modified. The modification consists in overlaying the sulfidic charge with either crushed pyrex glass, ground silicates (*e.g.*, feldspar), or quartz powder. Upon heating the silicates will melt and, because of the internal helium gas pressure, protect the refractory sulfide in a similar way as illustrated in Fig. 2E. Unfortunately, the viscosity of these melts decreases with decreasing melting points from pure SiO_2 towards alkaline silicates. Also these silica liquids will react with the Al_2O_3 DTA crucibles. In those cases, small crucibles consisting of either graphite (as can easily be prepared from arc electrodes), boron carbide, or perhaps boron nitride may be used.

The above DTA apparatus is situated in the Pulvermetallurgisches Laboratorium, Max-Planck-Institut für Metallforschung, Stuttgart-Büsnau. Its principal data are: a tungsten self-heating head which is inserted into a water-cooled autoclave filled with helium. DTA heating rates of 20 and 80 °C per min may be used for the high-temperature experiments. The measuring thermocouples are W/W-Re (20%). Calibration curves are referenced to: iron (Curie-point, α/γ, γ/δ, melting point), BeO (α/β), or gold (melting point = 1064 °C), cobalt (MP = 1493 °C), platinum (MP = 1769 °C), ruthenium (MP = 2350 °C); (cf. Moh *et al.*[5]).

Phase Identification

All reaction products were routinely examined at room temperature. Identification of phases was often possible before the transparent silica glass tubes were opened. Examination with a magnifying glass or binocular commonly indicated the degree of reaction and phases present. Color and luster (metallic or nonmetallic) often enabled phase identification. Textures due to sintering, recrystallization, or partial or complete melting were observed after quenching. Streak, hardness, specific gravity, and refractive indices were studied on the individual reaction products and compared in order to distinguish the coexisting phases.

Small polished sections of the reaction products were prepared for reflecting-light microscopy by using "Caulk Kadon" or "Technovit" as a mounting medium. Identification of the phases either was undertaken in air or was aided by using oil immersion. In polished sections, depending on the optical properties of the phases examined, identification is often possible when as little as 0.01% of a phase is present.

X-ray powder diffraction patterns were obtained with a Philips diffractometer, using Ni-filtered $Cu_{K\alpha}$ radiation or Mn-filtered $Fe_{K\alpha}$ radiation. For routine examinations scanning of the 10° to 60° 2 Θ region proved sufficient.

Binary Systems

Our discussion deals with sulfides formed from elements of the VI-B group. These are some of the most temperature-resistant sulfides, with the exception of CaS, thorium sulfides, and those of the rare earths. The present paper reviews the recent literature and supplements it to a certain extent (for literature sources, see[6–10]).

Generally the Cr–S system is more complicated, but with increasing atomic weight the number of stable sulfides decreases, while their temperature stability increases in the Mo–S, W–S system.

The Cr–S System

Of the VI–B group elements, chromium has the lowest melting point, 1875 ± 10 °C. The stable sulfides occur within a limited compositional range (nearly 50–60 at. % S) at intermediate temperatures for a high-temperature system. A total of seven phases has been reported, and the phase relations are fairly well established over a large temperature scale. Recently the system was investigated by El Goresy and Kullerud[11, 12] with respect to the occurrence of extraterrestrial chromium sulfides (*e.g.*, in meteorites). Modern laboratory techniques were employed, and a critical review of previous literature data was also included[12]. New experimental results and those of earlier work were presented in a schematic *T*-X diagram covering a temperature range from 300 to 1900 °C, as shown in Fig. 3. The dashed lines represent data of less accurate determination. The stable phases found by El Goresy and Kullerud[12] differ partly from compounds reported by Jellinek[13]. Fig. 3 shows metallic chromium coexisting with α-$Cr_{1.03}$ S at low temperatures. The α-phase can be observed with less than 39.61 wt. % S as an intergrowth with or exsolved from the Cr_{1-x} S phase, and with chromium. The Cr_{1-x} S phase (indicated in Fig. 3 as β) occurs in three different crystallographic modifications. Quenching experiments indicated:

1. β as a NiAs-type modification
2. a NiAs supercell hexagonal form
3. a low-temperature monoclinic form.

β-Cr_{1-x} S was found in all experiments[12] with less than 41.74 wt. % S. In quenching experiments containing 39.61 wt. % S or less, α-$Cr_{1.03}$ S has exsolved, as observed *via* reflected-light microscopy. In those experiments with less than 37.62 wt. % S, β-Cr_{1-x} S occurred together with α-$Cr_{1.03}$ S and Cr.

Hexagonal Cr_{1-x} S with a NiAs superstructure was obtained in experiments containing 42.04 to 43.53 wt. % S, whereas monoclinic Cr_{1-x} S occurred within the 43.53 to 44.79 wt. % S range, in coexistence with rhombohedral γ-$Cr_{2.1}$ S_3 solid solution.

As shown in Fig. 3, the compositional range of β narrows at higher temperatures and displays a singular point (approximately 1650 °C).

The rhombohedral γ-$Cr_{2.1}$ S_3 solid solution series has a stability field ranging in composition from 46.84 to 45.12 wt. % S, but its precise incongruent melting tem-

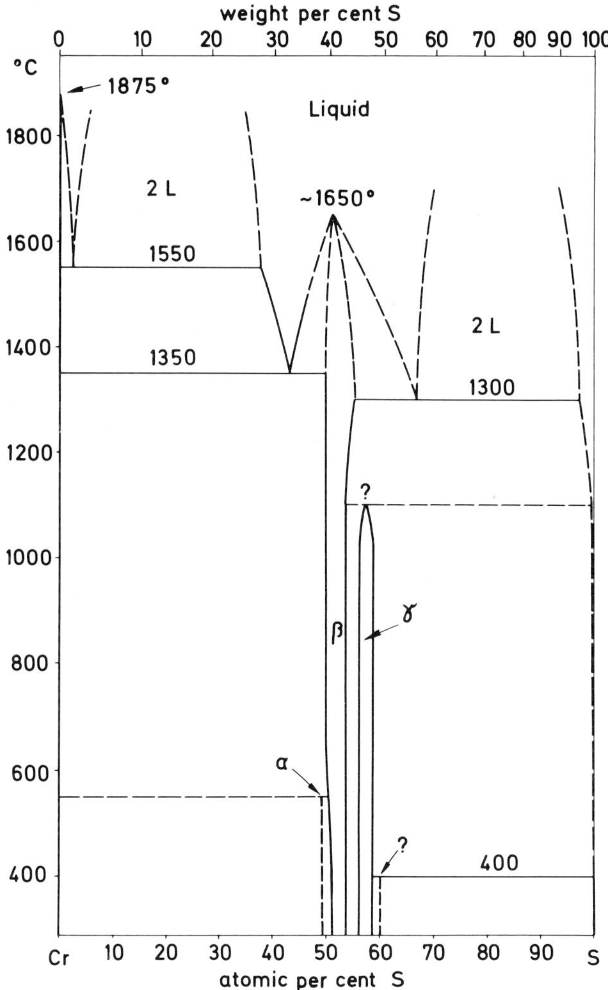

Fig. 3. The Cr–S system (partly schematic) after El Goresy and Kullerud[11]. Compositional data of the stable binary phases are given in the text

perature is unknown, as indicated by a dashed line. The breakdown products of the γ-phase are β and liquid S.

An additional sesquisulfide $Cr_{0.69}$ S reported to be of triclinic symmetry[13] was not encountered within the experimentation[12]. The liquidus relations of the chromium-sulfur system, as taken from literature data, display the Kullerud Type 1 of classification of binary sulfide or selenide systems[14]. This type of phase diagram is characterized by the occurrence of two regions of liquid immiscibility: one field with two immiscible liquids between metal and a congruently melting compound, and a second liquid immiscibility field in the region between the compound and sulfur or selenium respectively. As shown in Fig. 3, the first immiscible liquids field occurs above 1550 °C, ranging from 2.2 wt.% S (monotectic) to approximately 27.5 wt.% S.

The second two-liquid field above ~1300 °C was assumed to be between a monotectic of approximately 55 wt.% S and a composition close to sulfur (which is supercritical at this temperature). The binary eutectic lies at ~1350 °C with approximately 67.5 wt.% Cr and 32.5 wt.% S between metallic chromium and the β-phase.

The Mo–S System

The melting point of molybdenum, 2620 ± 10 °C, is rather high, only surpassed by a few other metals, namely osmium, tantalum, rhenium, and tungsten. The most important source of this much-desired metal is molybdenite (MoS_2) which is commonly related to acidic rocks and generally ascribed to pegmatitic-pneumatolytic origin. MoS_2 can occur as normal hexagonal molybdenite with a typical layered-structure, in rhombohedral form[15–18], and in another form reported as occurring naturally, but X-ray amorphous namely, jordisite[19]. Because of this polymorphism and because of controversial literature data about the occurrence of other sulfidic compounds, the binary Mo–S system has been subjected to repeated investigation. Nevertheless a phase diagram has not been presented, because of the high temperature stability of the phases concerned and the inadequate experimental techniques employed (for instance, the reported melting points for MoS_2 differ greatly, from 1185 to ~1800 °C, as cited in[6–8]).

In addition to the disulfide, MoS_2, a sesquisulfide, Mo_2S_3 or $Mo_{2.06}S_3$, and Mo_3S_4, Mo_2S_5, MoS_3, MoS_4, and also a polysulfide, $MoS_3 \cdot S_x$, have been reported. Most of the references are given in Gmelin's Handbook[20], and more recent data are compiled in[6–8]. Of special interest is the Mo_3S_4 compound described by Chevrel *et al.*[21]. The Mo_3S_4 composition could not be synthesized directly from the elements or binary compounds, thus previously synthesized ternary phases, $M_xMo_3S_4$, were leached with inorganic acids to produce the Mo_3S_4 composition.

Only two compounds of the binary Mo–S system can be synthesized directly from the pure elements: the sesquisulfide with a composition of $Mo_{2.06}S_3$, and hexagonal MoS_2[22]; no other phases were obtained by dry sulfide experimentation. When using molybdenum powder and sulfur as starting materials, the $Mo_{2.06}S_3$ phase appears above 610 ± 5 °C, whereas below this temperature Mo and MoS_2 coexist. Heating experiments of 1 month show that, after its formation, $Mo_{2.06}S_3$ will not decompose at or below 600 °C. The sesquisulfide can be quenched to room temperature where it remains metastable. Due to the sluggish reaction rates of the system, previously synthesized MoS_2 and molybdenum did not react to produce any sesquisulfide at 650 °C in the course of 30 days. With increasing temperatures the reaction kinetics will slowly increase; thus, equilibrium assemblages can be attained at 900 or 1000 °C in less than 1 week. Morimoto and Kullerud[22] have determined the exact composition of the sesquisulfide, $Mo_{2.06}S_3$, at 935 °C; this deviates slightly from the stoichiometric ratio. The sesquisulfide was first reported to crystallize in a tetragonal structure with a = 8.6 Å, c = 10.9 Å[23]. Other literature data give monoclinic symmetry, with a = 8.6335 Å, b = 3.208 Å, c = 6.902 Å, β = 102°43′[24].

Normally, MoS_2 appears in its typical hexagonal form. A rhombohedral modification was synthesized at about 900 °C[15], but the structural differences between the hexagonal and the rhombohedral forms of MoS_2 can be explained by assuming different stacking orders of the S–Mo–S layers[25, 26]. X-ray powder diffraction patterns of MoS_2 synthesized at temperatures below 900 °C give broad peaks, and the lower the temperature of synthesis the broader are the peaks. These poorly defined peaks do not fit exactly with those of the hexagonal form, nor with those of the rhombohedral form. On heating above 900 °C, all X-ray patterns show sharp peaks of the hexagonal modification[22]. A transformation can be observed in heating experiments to 1000 °C or above with natural rhombohedral molybdenite-3R or with X-ray amorphous jordisite, which is rarely found in low-temperature deposits. At high temperatures only the hexagonal form of molybdenite can be attained. Once synthesized, hexagonal MoS_2 does not change to the rhombohedral modification or to any intermediate form, even at low temperatures or after prolonged heating.

By employing high-temperature techniques for dry sulfide experimentation as described above, it was possible to study the temperature stability relations of the phases and binary reactions within the Mo–S system. Preliminary results were reported by Moh *et al.*[27]. The sesquisulfide was found to melt incongruently to a metal-rich liquid and solid hexagonal MoS_2. The first reported incongruent melting temperature of ~1740 °C seems to be a little too high. DTA experiments commonly show superheating and supercooling as well, which complicates the exact determination of the incongruent melting temperature. Repeated heating and cooling of the same DTA run yielded lower temperatures, due to some loss of sulfur vapor, which would result in a steadily shifting composition of the charge towards the binary eutectic (cf. Fig. 7). Additional heating experiments (compare the techniques illustrated in Fig. 2) yield a breakdown temperature of the sesquisulfide slightly below the melting temperature of silica glass. Figure 4 shows a macrophotograph of the sulfidic charge of such an experiment in which the silica glass tube has been nearly but not completely molten, whereas the charge definitely produced MoS_2 crystals. Figure 5 again shows a polished section of another regulus of the same heating experiment: stoichiometric $Mo_{2.06}S_3$ was heated above its incongruent melting point, and during the subsequent cooling period most of the sulfide has again been crystallized to a coarse-grained matrix with some molybdenum metal inclusions and remaining MoS_2 crystals on the edges, displaying disequilibrium conditions due to fast cooling.

The eutectic temperature of the system is 1610 ± 15 °C, whereas the composition of the binary eutectic has been revised recently by Moh[4] to 42 at. % sulfur and 58 at. % molybdenum with an accuracy of ± 1%, as illustrated in Fig. 6A, B, C. For exact determination of the eutectic composition, a series of quenching experiments was performed with charges of coexisting Mo and $Mo_{2.06}S_3$ (previously synthesized from the elements) and then chilled from 1630 °C.

A number of additional DTA experiments were undertaken with various compositions within the binary system up to 66.6 at. % S (= MoS_2 composition). In Fig. 7 a phase diagram is shown in which all results are incorporated. With respect to the above-mentioned classification of sulfide systems[14], the Mo–S system, as well as the Cr–S system, exhibits Type 1: two regions of immiscible liquids; one field of liquid immiscibility in the metal-rich portion at high temperatures, and a second two-liquid field in the sulfur-rich region beyond MoS_2 which is not shown in Fig. 7.

Fig. 4. A macroscopic picture of an incongruently melted molybdenum sesquisulfide regulus. A melting reaction has formed molybdenite (MoS_2) at the edges. The initial charge (pure $Mo_{2.06}S_3$) was heated to approx. 1700 °C and chilled within 30 min to room temperature. Regulus size 3.5 mm diameter. Photo Udubasa

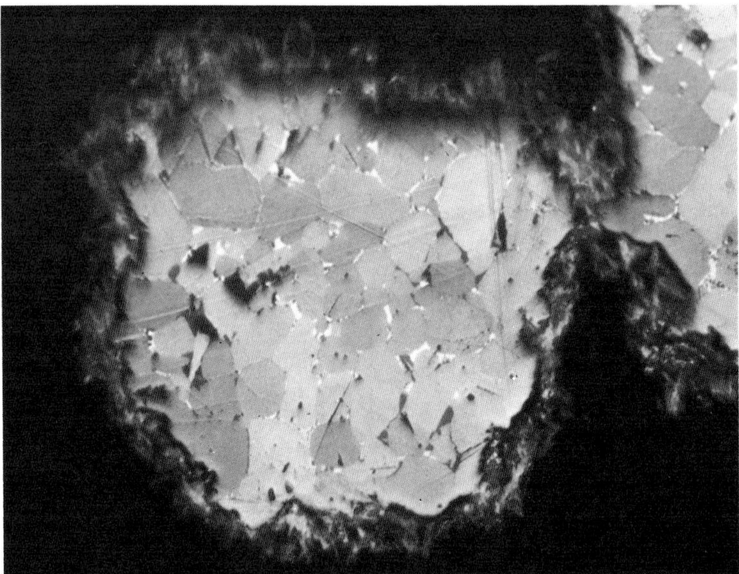

Fig. 5. Reflected-light micrograph of an incongruently melted $Mo_{2.06}S_3$ sample, rapidly cooled from 1700 °C, displaying disequilibrium assemblages. During solidification, molybdenum sesquisulfide has again formed as a coarse-grained, distinctly pleochroic matrix (*different shades of grey*) with some inclusions of metallic molybdenum (*white*) and feathered MoS_2 crystal aggregates on the edges. Regulus diameter approx. 3 mm; oil immersion. Photo Udubasa

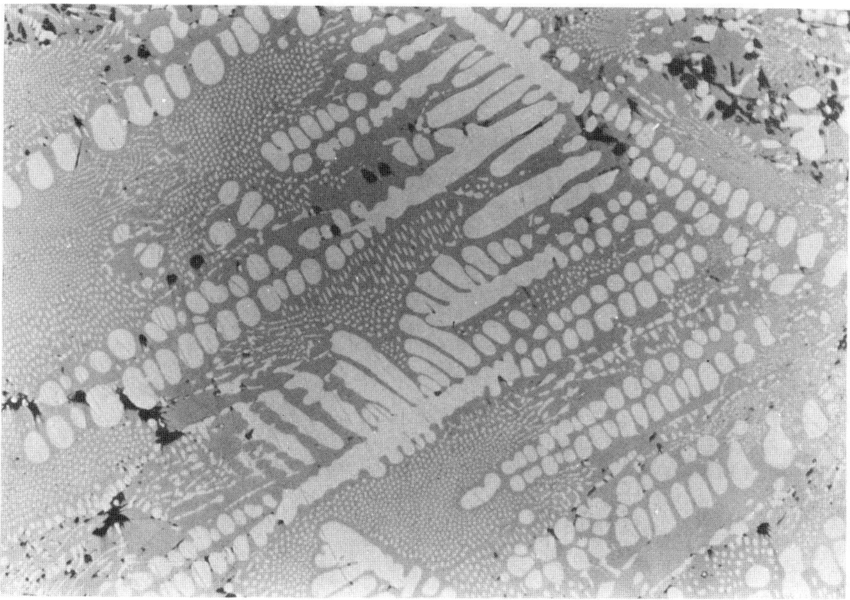

Fig. 6 A. Excess molybdenum metal (*light*) in an eutectic matrix of the Mo–S system (40 at. % S), chilled from 1630 °C; x 400, in air

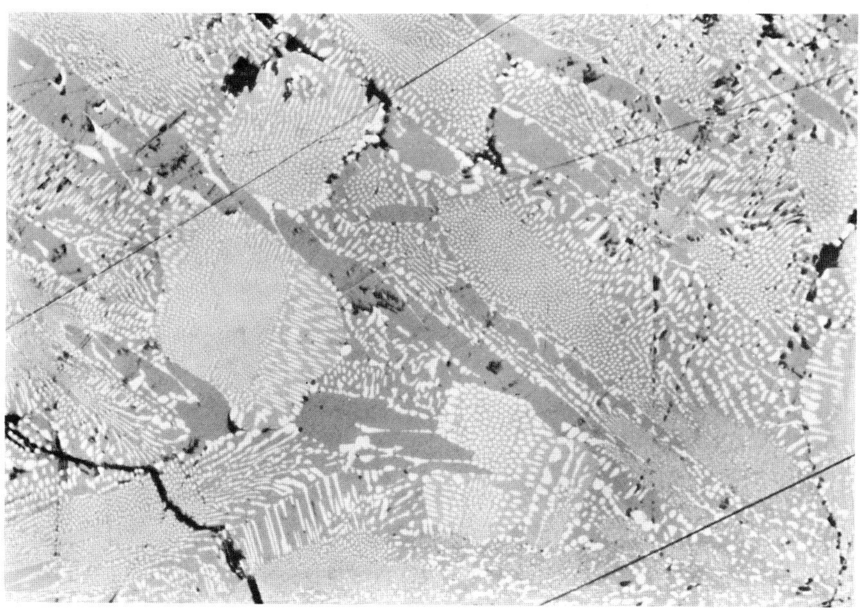

Fig. 6 B. Microphotograph showing a composition of 57.1 at. % Mo, 42.9 at % S, of nearly the binary Mo–S eutectic, chilled from 1630 °C; x 400, in air

Fig. 6 C. Crystals of molybdenum sesquisulfide (46.5 at. % S) displaying distinct cleavage, partly with liquid inclusions, floating in the binary Mo–S eutectic melt, chilled from 1630 °C; x 400, in air

Recent high-temperature DTA experiments yielded the expected region of two liquids in the metal-rich portion above ~ 1950 °C[28], but neither the exact composition of the monotectic (liquid 1) nor that of the coexisting liquid 2 is known. The DTA results of a successful experiment with a mixture of Mo and $Mo_{2.06}S_3$, of ~ 18 at.% S composition, definitely showed the eutectic temperature to be 1620 °C, the monotectic temperature to be 1950 °C, and a third effect at ~ 2050 °C, which may indicate a homogeneous liquid above that temperature.

Differential thermal analyses carried out with pure $Mo_{2.06}S_3$ and with mixtures of $Mo_{2.06}S_3$ + Mo or with $Mo_{2.06}S_3$ + MoS_2 resulted in distinct thermal effects on both heating and cooling at ~ 1560 °C which were interpreted as α-β phase transition for the sesquisulfide[27, 28], as indicated in Fig. 7. Continuous heating at heating rates of or less than 20 °C per min[28] shows the incongruent melting point of the high-temperature α-modification of $Mo_{2.06}S_3$ to be slightly below 1700 °C. No thermal effects of β-$Mo_{2.06}S_3$ were obtained at low temperatures (~ 610 °C), which is in agreement with the sluggish reaction kinetics reported above[22].

No thermal effects were recorded for MoS_2 in DTA experiments performed up to ~ 1800 °C. Some heating experiments were carried out (employing the techniques illustrated in Fig. 2) to obtain information about the thermal stability relations of MoS_2. The solid sulfidic charge remained stable when shielded by the "molten" silica glass reaction vessel in equilibrium with the outside atmospheric pressure at ~ 1750 °C. Upon heating, MoS_2 began to decompose and vapor was lost because of the increasing partial sulfur pressure of the charge. In another experiment, after a

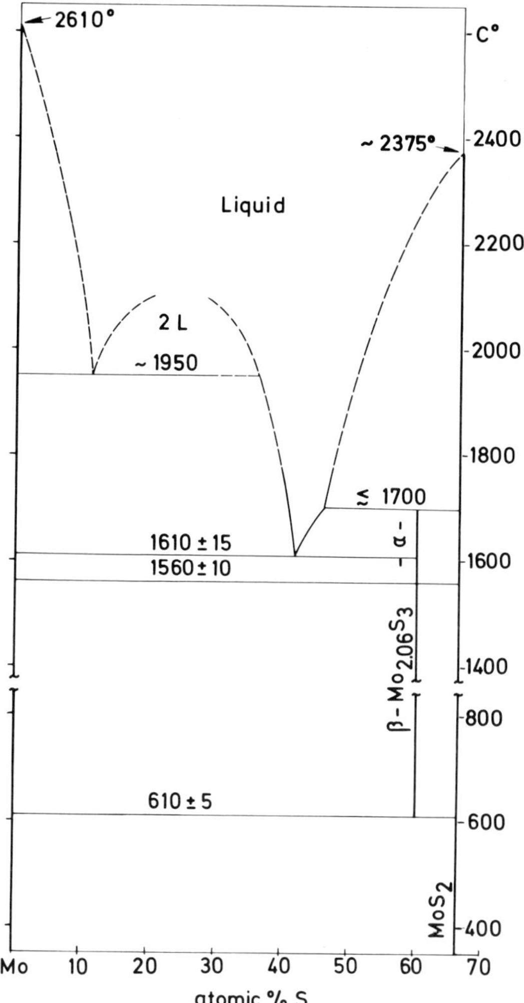

Fig. 7. The condensed Mo–S system from 0 – 66.6 at % sulfur (= MoS_2 composition). Vapor is present as a stable phase in all assemblages, and the vapor pressure is not constant

short heating period up to 1800 °C, the subsequent examination of the chilled product indicated that more than half of the MoS_2 had decomposed; thus, liquid and relics of MoS_2 existed at this reaction temperature. These experimental results are in good agreement with other findings and speculations by Cannon[29]: the melting temperature of MoS_2 is assumed to be ~2375 °C (at high equilibrium pressure). It should be mentioned that recent high-temperature experiments with MoS_2 carried out in a plasma-jet melting apparatus[30] under argon pressure indicate that molybdenite is stable in equilibrium with a metal-rich liquid above 2000 °C, as demonstrated in Fig. 8. However, exact temperature determination has not yet been possible.

The phase diagram in Fig. 7 does not include the sulfur-rich portion. X-ray amorphous trisulfide of approximately MoS_3 composition cannot be synthesized in the pure dry system[31]; however, MoS_3 is reported to occur as a mineral in nature[32]. Syntheses are possible either as precipitate in acidic aqueous solutions or

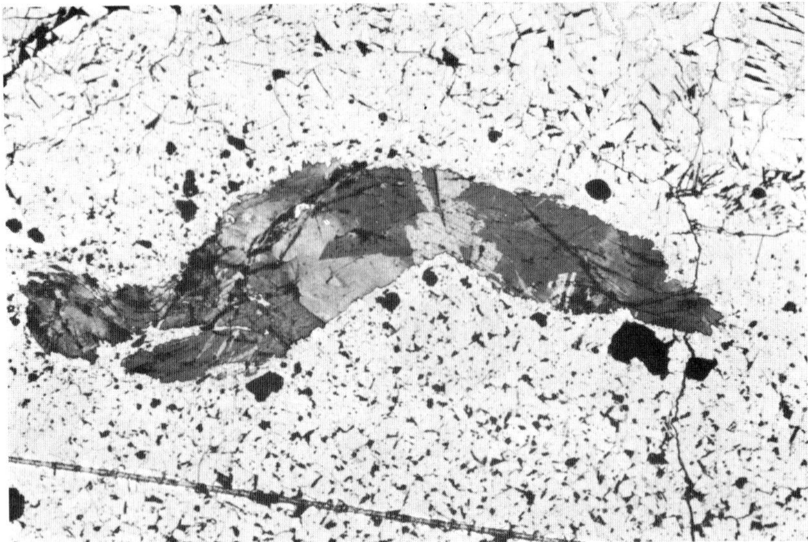

Fig. 8. Partly corroded molybdenite (MoS_2) fragments, strongly pleochroic, floating in a metal-rich sulfidic melt. Heated in a plasma-jet melting apparatus, chilled from above 2000 °C; x 2000, oil immersion

by a dry breakdown reaction of $(NH_4)_2MoS_4$ between 190 and 200 °C. The trisulfide obtained will decompose on heating above ~250 °C to liquid sulfur and molybdenum disulfide. Its decomposition is as follows: first, X-ray amorphous disulfide appears; with increasing temperatures it becomes rhombohedral, and, in turn, transforms into the hexagonal modification of MoS_2 at 900 to 1000 °C[33)].

With regard to the above-mentioned classification of sulfide systems, a region of two immiscible liquids should be assumed at high temperatures, above the critical temperature of sulfur. These liquidus reactions are part of present investigations, the results of which will be published at a later date.

The W–S System

With ~3410 °C tungsten has the highest melting point of all metals and is only surpassed by carbon. Tantalum and hafnium carbides, however, have even higher melting points. When compared with molybdenum, tungsten has a much higher affinity to oxygen and a lower affinity to sulfur. The most common tungsten minerals are scheelite ($CaWO_4$) and wolframite ($FeWO_4$). Tungstenite (WS_2) is a very rare mineral; but owing to its similarity to molybdenite it has been mistaken for molybdenite[18,19,34)].

The disulfide (WS_2) is the only stable compound which can be synthesized directly from the elements. Previous literature data are compiled in Refs.[6–8)]. The W–S system presented in Fig. 9 is based on literature data and schematically illustrates possible phase relations. Normally, WS_2 crystallizes hexagonally and has the same structure as molybdenite; under certain conditions, however, a rhombohedral modification of WS_2 can be prepared. The trisulfide WS_3 cannot be synthesized directly from the elements, but, as is the case for MoS_3, an X-ray amorphous phase has been reported; this is indictated in Fig. 9.

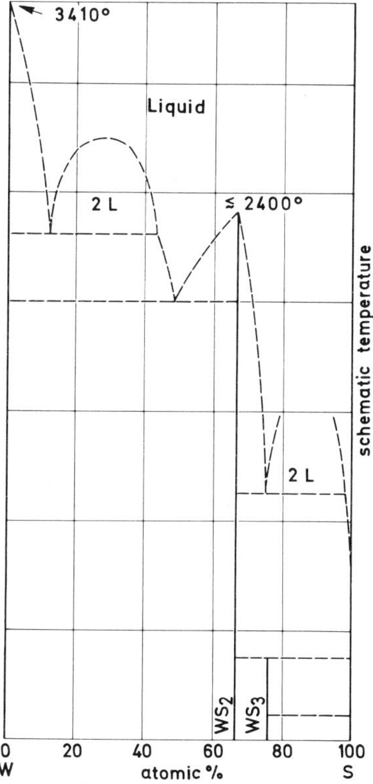

Fig. 9. Schematic illustration of the W–S system, based on literature data

When WS_2 is prepared below 700 °C, a disordered crystal structure characterized by a different stacking sequence is produced. Appreciable ordering will occur on heating to higher temperatures[35]. Structural properties of hexagonal and rhombohedral MoS_2 and WS_2 are illustrated, including the various stacking faults with respect to the X-ray diffraction pattern[25]. The rhombohedral modification was obtained by experimentation in an alkali carbonate flux. The reported compositional range for tungsten disulfide ($WS_{2\pm x}$) was not confirmed in the 800 to 1000 °C range, even by weighing the stoichiometric elements of high purity directly into the silica glass tubes (cf.[2]). In all cases WS_2 was found to be a stoichiometric compound. Weighings with lower sulfur contents yielded small relics of metallic inclusions in the tungstenite produced. However, a verification of the reported WS_2 nonstoichiometry at lower temperatures was not performed. The formation of nonstoichiometric disulfide may occur metastably at low temperatures, for instance, if the nonschoichiometric disulfide is formed from the break-down of WS_3 under different conditions not belonging to the pure dry system. Thus, a slight oxidation of WS_2

can hardly be recognized by wet chemical analytical procedures and oxidation could cause modifications to the diffraction pattern.

Literature data show the impossibility of preparing a pure WS_3 phase. It always shows distinct signs of oxidation (3–4% oxygen) and/or NH_4Cl contamination of 8% or more[26, 36]. On heating, the trisulfide will melt incongruently to WS_2 and liquid S within the 270 to 500 °C range[8], depending on the following conditions: vacuum, atmospheric pressure, equilibrium with excess sulfur.

WS_2 thermal stability is still obscure. At temperatures below ~ 1150 °C no signs of a decomposition were noticed and WS_2 was found to sublimate slowly but in weighable amounts, *e. g.*, in a zone from 1000 to 900 °C. The product consisted of a well-crystallized hexagonal WS_2[36]. Heated in vacuum, WS_2 starts to decompose at 1200 °C. The congruent melting point of WS_2 lies above 1800 °C. However, when compared with the MoS_2 melting relations[29], WS_2 should melt in the 2400 °C range with a high equilibrium sulfur vapor pressure.

In recent heating experiments, pure synthetic WS_2 was heated up to 1800 to 1850 °C, using the method illustrated in Fig. 2. WS_2 did not decompose, indicating a partial sulfur vapor pressure of less than 1 atmosphere. At ~1850 °C or above, WS_2 breaks down. In one experiment at 1900 °C from the original sulfidic charge only metallic tungsten remained. After a careful microscopic and X-ray examination, this tungsten sponge showed no indications of carbide formation caused by the possible contamination with the graphite crucible. No sulfur was found, only metallic tungsten.

Like the other systems above, the W–S system can be treated according to Kullerud's classification scheme for binary sulfides[14]. Two regions of liquid immiscibility should occur at high temperatures under equilibrium pressure in both the metal-rich and the sulfur-rich portions of the W–S system. This is incorporated in the schematic diagram shown in Fig. 9.

Ternary Systems

The previously studied ternary sulfide systems included chromium, molybdenum, and/or tungsten with the added elements iron, copper etc. While other isotherms may be available, only that of 700 °C will be depicted. Higher temperature equilibria are deduced from these and their reactions are presented. Lower temperature phase equilibria of geologic importance are referenced.

The Mo–W–S System

Of the Cr–Mo–S, Cr–W–S, and Mo–W–S ternary systems, only Mo–W–S has been studied from 500–1000 °C, by Moh and Udubasa[37]. Details of the binary systems Mo–S and W–S are given above. The metallic system Mo–W is characterized by a complete series of solid solutions between the endmembers with continuous liquidus and solidus relations[38]. Complete solid solution series were also found to occur between the two hexagonal molybdenum and tungsten disulfides. At 1000 °C this complete series coexists with liquid sulfur and metallic tungsten. The sesquisulfide $Mo_{2.06}S_3$ is stable with the complete intermetallic solid solution series. Thus, in the central part of the ternary system exists the univariant phase assemblage $Mo_{2.06}S_3$ + MoS_2 + W + vapor.

By comparing sulfur activity data at or below 800 °C from the literature, the ternary phase relations will change. The sesquisulfide does not coexist any longer with the complete intermetallic Mo–W series. This divariant region narrows with diminishing temperatures towards the Mo-rich portion. Simultaneously MoS_2 becomes stable not only with metallic tungsten but also with an increasing part of the intermetallic solid solution series beginning with tungsten and extending gradually towards Mo-richer endmembers with decreasing temperatures. The approximate phase relations at ~700 °C are illustrated in Fig. 10. Finally below 610 ± 5 °C the $Mo_{2.06}S_3$

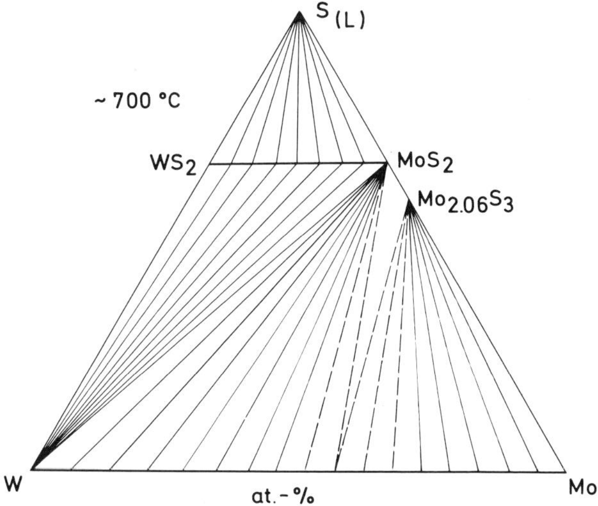

Fig. 10. The Mo–W–S system at approx. 700 °C

compound is no longer stable in the Mo–S binary system, and MoS_2 coexists with the complete intermetallic molybdenum-tungsten solid solution series, whereas pure tungsten is in equilibrium with the complete (Mo, S)S_2 series.

Low-temperature relations and geologic applications are discussed by Moh and Udubasa[37]. Naturally occurring members of the MoS_2–WS_2 solid solution series are reported in Refs.[17–19].

Related Ternary Sulfide Systems

Some of the following ternary sulfide systems are briefly discussed with respect to related mineral occurrences. In addition to the 700 °C isotherms and the temperature-dependent reactions, the most recent experimental studies are included and illustrated by micrographs.

The Fe–Cr–S System

El Goresy and Kullerud[12)] correlated the physicochemical conditions of mineral formations in meteorites to the Fe–Cr–S system. The Cr–S subsystem is displayed in Fig. 3.

In the Fe–S binary system, at 700 °C, two binary phases are stable: pyrrhotite ($Fe_{1-x}S$) with a hexagonal NiAs-type structure, and pyrite (stoichiometric FeS_2), a P_{A3}-structure-type mineral. Pyrite melts incongruently at 743 °C to pyrrhotite and liquid sulfur[40)], whereas the solid solution composition of pyrrhotite narrows with increasing temperature and a singular point occurs at 1192 °C with an $Fe_{0.92}S$ composition containing ~52 at.% S. Troilite, a meteoritic mineral is stoichiometric FeS. Power and Fine[41)] include references for low temperature within the Fe–S system.

The metallic Cr–Fe system has been investigated in detail; references are summarized in[6, 7)].

The phase relations in the ternary Fe–Cr–S system are influenced by the extended ternary monosulfide solid solution series of the binary endmembers: $Fe_{1-x}S$ and $Cr_{1-x}S$ (see Fig. 11). This series increases rapidly with increasing temperatures

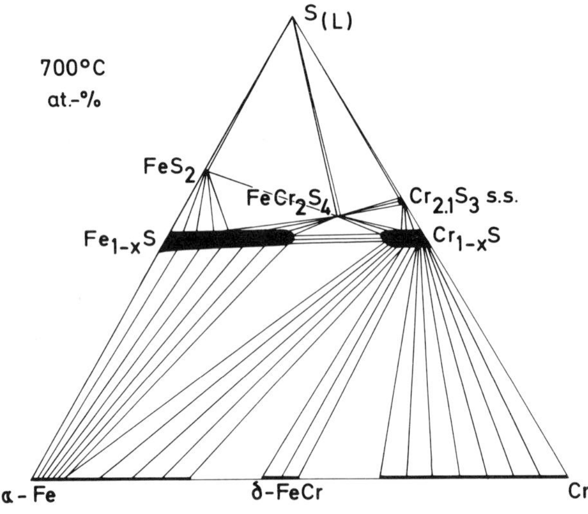

Fig. 11. The 700 °C isotherm of the Fe–Cr–S system after El Goresy and Kullerud[12]

and becomes complete slightly above 700 °C. In the metal-rich portion of the system both monosulfide mixed crystal rows coexist with the metallic α-Fe solid solution, whereas the more Cr-rich members of the $(Cr, Fe)_{1-x}S$ series are also stable with δ-FeCr and with all of the chromium solid solution series.

The common meteoritic mineral daubreelite, $FeCr_2S_4$, a sulfur spinel, occurs as a ternary phase in the sulfur-rich portion of the system. It coexists at 700 °C with $Cr_{2.1}S_3$, with members of both monosulfide series, with pyrite (FeS_2) and with liquid sulfur. However, at lower temperatures, the ternary extension of both monosulfide rows will decrease, so that, *e.g.* at 600 °C, daubreelite is also stable with metallic α-Fe from the invariant reaction:

$$(Fe, Cr)_{1-x}S + (Cr, Fe)_{1-x}S + \text{vapor} \rightleftharpoons FeCr_2S_4 + \alpha\text{-Fe}$$

which takes place at 650 ± 50 °C[12].

Figure 12 shows daubreelite exsolution lamellae in troilite; the sample is from the Mundrabilla meteorite described by Ramdohr[42].

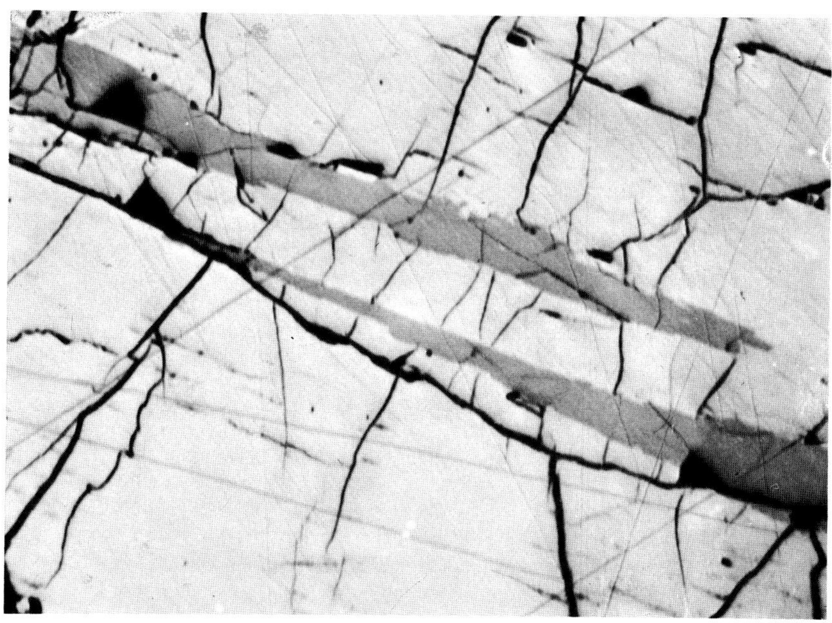

Fig. 12. Coarse daubreelite lamellae ($FeCr_2S_4$) and a finely dispersed second generation exsolved from troilite (FeS) parallel to (0001); cleavage and cracks of shrinkage are visible. Mundrabilla meteorite, oil immersion, x 400

High pressure/high temperature quenching and DTA experiments indicated that cubic α-daubreelite has a monoclinic β-polymorph at pressures of more than 14 kb; the reactions were discussed on the basis of a *p-T* diagram[43].

The Fe–Mo–S System

With a view to making a complete study of the quaternary Fe–Cu–Mo–S system, the Cu–Mo–S system was investigated; it is discussed below. Simultaneously, the phase relations within the ternary Fe–Mo–S system were considered between 500 and 750 °C and extended partly to 1000 °C[44–46]. In comparison with the Fe–Cr–S ternary phase relations, the Fe–Mo–S system and all the following ones differ greatly.

Figure 13 represents the Fe–Mo–S system at 700 °C. The metallic Fe–Mo system has frequently been investigated in great detail; references are summarized in[5–7]. At 700 °C only α-Fe, the hexagonal phase Fe_3Mo_2, and Mo occur, all with limited solid solution (compare the intermetallic binary solid solubility range in[46]).

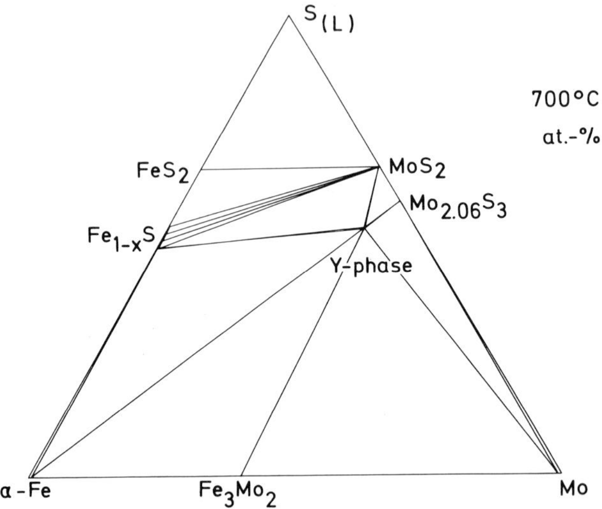

Fig. 13. The 700 °C isotherm of the Fe–Mo–S system

As shown in Fig. 13, molybdenite (MoS_2) coexists at 700 °C with FeS_2 and $Fe_{1-x}S$, but not with α-Fe. Molybdenite with its layered structure is commonly used as a lubricant for steel alloy bearings up to elevated temperatures. The phase equilibria (Fig. 13) clearly show the unsuitability of MoS_2 lubricants at 700 °C: iron and MoS_2 will react to form FeS and a ternary compound, which was called Y-phase by Grover and Moh[44]. The occurrence of the ternary compound influences the phase relations within the metal-rich portion of the Fe–Mo–S system. It crystallizes rhombohedrally[47] with a composition of approximately $FeMo_4S_6$ coexisting with MoS_2, $Mo_{2.06}S_3$, metallic molybdenum, Fe_3Mo_2, α-Fe, and with stoichiometric FeS of the $Fe_{1-x}S$ solid solution series. Y-phase coexistence with FeS prevents α-Fe and MoS_2 from existing together at 700 °C. The Y-phase can be synthesized directly from the elements at 700 °C, or more rapidly at 1000 °C in a reaction between iron powder and molybdenum sesquisulfide:

$$Fe + 2\ Mo_{2.06}S_3 \longrightarrow FeMo_{4.12}S_6$$

producing a slightly Mo-rich Y-phase. The Y-phase is not a stoichiometric ternary compound, occurring with a considerable solid solution width as a function of the formation temperature. Comprehensive reflected-light microscopy and X-ray diffraction investigations of quenched products of various compositions indicate a solid solution width of several per cent, with a reference composition of $FeMo_4S_6$ or $FeMo_{4,12}S_6$ respectively[4, 5, 46]. The solid solution width of the Y-phase increases with temperature and shows many similarities with the analogous X-phase ($\sim CuMo_2S_3$) as found in the Cu–Mo–S system; see Fig. 19.

A great variety of formulas for this phase can be found, *e.g.*[48–50]. Some of these data are dubious, others are only correct in respect of the temperature-dependent solid solution range of the Y-phase. However, the Y-phase should be described as a compound with a Fe : Mo ratio of 1 : 4, as this represents the only formula valid for the temperature stability range from slightly below 800 °C up to its melting point. The Y-phase can only be synthesized above 535 ± 15 °C[46]; it is quenchable and probably remains metastable at room temperature.

The lattice parameters are:

$a_{rh} = 6.47$ Å, $\alpha_{rh} = 94°39'$, $V = 268.7$ Å^3 (rhombohedral setting)
or $a_h = 9.52$ Å, $c_h = 10.27$ Å, with an axial ratio
$a_h/c_h = 0.927$ (hexagonal setting)[47].

The phase relations in the system are as follows:

500 °C	Since the ternary phase is not stable, MoS_2 coexists with all phases of the system (molybdenum solid solution, Fe_3Mo_2, α-Fe solid solution, $Fe_{1-x}S$, FeS_2, and liquid sulfur). With increasing temperatures a number of invariant reactions take place in the presence of a vapor phase (v). (For simplicity, in the formulas below only the way of the reactions is given and the balanced equations are omitted.)
535 ± 15 °C	The Y-phase forms as a ternary reaction; $MoS_2 + Mo + Fe_3Mo_2 \rightleftharpoons$ Y-phase + v
558 ± 9 °C	A switch of tie lines takes place; $MoS_2 + Fe_3Mo_2 \rightleftharpoons$ Fe + Y-phase + v
610 ± 5 °C	$Mo_{2.06}S_3$ appears in the Mo–S binary system between MoS_2 and Mo, and above it coexists also stably with the Y-phase.
677 ± 3 °C	Molybdenite (MoS_2) becomes unstable with metallic iron; $MoS_2 + Fe \rightleftharpoons FeS$ + Y-phase + v
726 ± 3 °C	A ternary monotectic forms in the sulfur-rich portion of the system; $MoS_2 + FeS_2 + S_{(liquid)} \rightleftharpoons$ ternary monotectic$_{(liquid)}$ + v

~735 °C	Molybdenite (MoS_2) becomes unstable with pyrite (FeS_2); $MoS_2 + FeS_2 \rightleftharpoons$ ternary monotectic$_{(liquid)}$ + $Fe_{1-x}S$ + v
743 °C	In the binary Fe–S system pyrite melts incongruently to liquid sulfur and pyrrhotite; $FeS_2 \rightleftharpoons Fe_{1-x}S + S_{(liquid)}$ + v
~973 °C	The ternary eutectic occurs in the metal-rich portion of the Fe–Mo–S system; Fe + FeS + Y-phase $\rightleftharpoons$ ternary eutectic$_{(liquid)}$ + v
988 °C	The ternary eutectic melt reaches the binary Fe–S system, the binary eutectic occurs; Fe + FeS $\rightleftharpoons$ liquid + v.

With small restrictions, the invariant reaction at 677 ± 3 °C (*i.e.*, the disappearance of the tie line between Fe and MoS_2) can be deduced from the intersection of curves 2 and 3 of the sulfur activity data shown in Fig. 17; however, the Y-phase is not indicated in that figure.

At even higher temperatures, various binary and ternary reactions occur, but only a few, which occur in the metal-rich portion, will be mentioned here. The Y-phase still appears in its low-temperature γ-modification. On further heating:

1100 ± 35 °C	the γ-modification inverts into the β-polymorph
1575 ± 15 °C	a high-temperature, probably triclinic α-form[47] becomes stable. It was found to melt at ~ 1700 °C
1700 ± 35 °C	as reported by Moh *et al.*[5].

Polished sections of charges of the first melting and DTA experiments indicated that the Y-phase melts incongruently[4, 47]. In most of the cases stoichiometric $FeMo_{4.12}S_6$ or $FeMo_4S_6$ were used as starting materials. A number of recent high-temperature experiments performed according to the techniques illustrated in Figs. 1 and 2 to minimize vapor loss, yielded either only slight decomposition or probably a congruent melting with a singular point at 1700 ± 35 °C and a composition of approximately 8.8 wt.% Fe, 60.2 wt.% Mo, and 31 wt.% S, as calculated within the experimental errors of the techniques employed. In microprobe analysis, the composition of the Y-phase near its melting temperature was found to be identical or almost identical with the chilled melt. Therefore, it cannot be determined whether the Y-phase melts incongruently or congruently at 1700 ± 35 °C. Thus, the following reactions are possible:

incongruent melting
1. Y-phase $\rightleftharpoons$ liquid + Mo + vapor
2. Y-phase $\rightleftharpoons$ liquid + Mo + vapor + trace amounts of a second FeS-rich melt

congruent melting
3. Y-phase $\rightleftharpoons$ liquid + vapor

At very low temperatures the Y-phase has the approximate formula $FeMo_{4.12}S_6$ and the solid solubility increases directly with temperature. At ~973 °C, which is the temperature of the ternary eutectic, the Y-phase has its Fe-richest composition ($Fe_{1.11}Mo_{4.03}S_6$).

Above this temperature the Y-phase coexists with an Fe-rich liquid which reduces its solid solution width. The solid solubility increases a little towards sulfur and the binary Mo–S system up to the appearance of melt at ~1500 °C (see Fig. 7). Results of various heating experiments show that the Y-phase of $FeMo_{4.12}S_6$ composition (or Mo-richer) will exsolve molybdenum metal at high temperatures (or MoS_2 at high sulfur fugacity). The quenched molten products from above 1700 °C indicate the coexistence of a sulfidic liquid and molybdenum. Fe-rich compositions of the Y-phase (stable below 973 °C) show that at high temperatures an FeS-rich liquid exsolves and separates. Microprobe analyses of products quenched from 1650 °C show that the FeS phase has dissolved up to 2.45 wt.% Mo.

At even higher temperatures, *e.g.*, 1800 °C or above, the partial sulfur pressure of the molten Y-phase approaches or exceeds 1 atmosphere, and a vapor loss is possible, especially for prolonged or repeated heatings. The quenched products commonly consist of a sulfidic matrix with increased amounts of metallic molybdenum, as shown in Fig. 14.

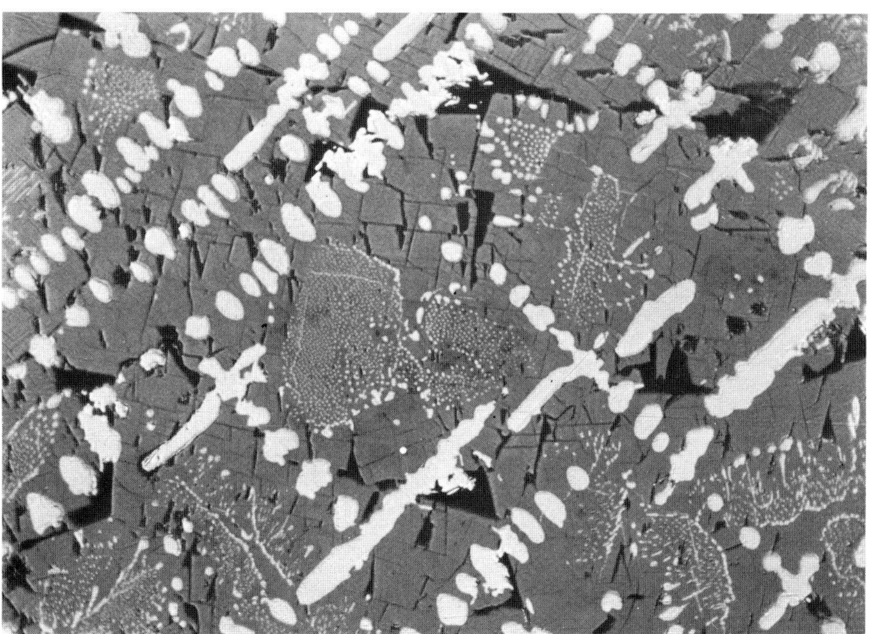

Fig. 14. Reflected-light micrograph of the breakdown products of $FeMo_{4.12}S_6$, melting point 1700 ± 35 °C. The charge was heated above 1800 °C and then rapidly cooled to room temperature. Owing to loss of vapor, metallic molybdenum (*white*) has exsolved in a skeletonlike texture of two generations: a coarse fragmented network and a second, finely distributed exsolution within the cleaved sulfidic matrix (*grey*); oil immersion, x 630

The phase relations in the metal-rich portion of the Fe-Mo–S system at ~1800 °C are as follows. Along the Fe–S side and within a large part of the Fe–Mo binary region exists a homogeneous liquid. The liquid phase boundary of the whole Fe-corner is connected by tie lines with molybdenum metal solid solution. In the Mo–S binary system (cf. Fig. 7) liquid occurs between 38 and 48 at.% S which extends into the ternary diagram up to or beyond Y-phase composition, as is shown by microprobe analysis[4)] of the quenched products. It was found by microprobe analysis that liquid immiscibility occurs between this molybdenum-rich sulfidic melt and a melt of nearly FeS composition; both liquids coexist with molybdenum solid solution containing 1% Fe. Figure 15 shows a light gray matrix of a quenched Mo-rich sulfidic melt with chilled liquid inclusions (slightly darker) of approximate FeS composition, and various inclusions of molybdenum.

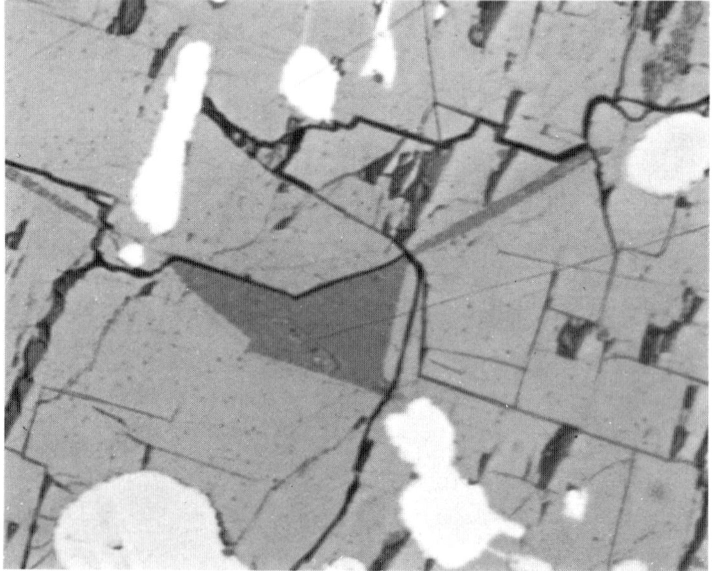

Fig. 15. Groundmass of a solidified Mo-rich sulfidic melt (*light grey*) with liquid inclusions of approximately FeS composition (*dark grey*) and exsolved molybdenum (*white*); oil immersion, x 2500

The sulfur spinel, $FeMo_2S_4$[48, 49)], was not found during experimentation. However, it is interesting to note that a real oxygen spinel was found, $FeMo_2O_4$, indicating trivalent molybdenum[4)].

The reported mineral femolite, $FeMo_5S_{11}$[51)], probably stable at low temperatures, could not be synthesized in the pure dry system[46)].

The Fe–W–S System

The Fe–W–S system below 743 °C (the breakdown temperature of pure pyrite[40)]) was investigated by Stemprok[52)] with regard to its geologic significance. Some reactions were undertaken to extend the system to 1000 °C. These experimental results were generally identical with those of an earlier study by Vogel and Weizenkorn[53], who concentrated on the metallurgic aspect.

Stemprok[52] discussed the phase relations from sulfur activity literature data, as shown in Fig. 16. The affinity of various metals to sulfur and their distribution in

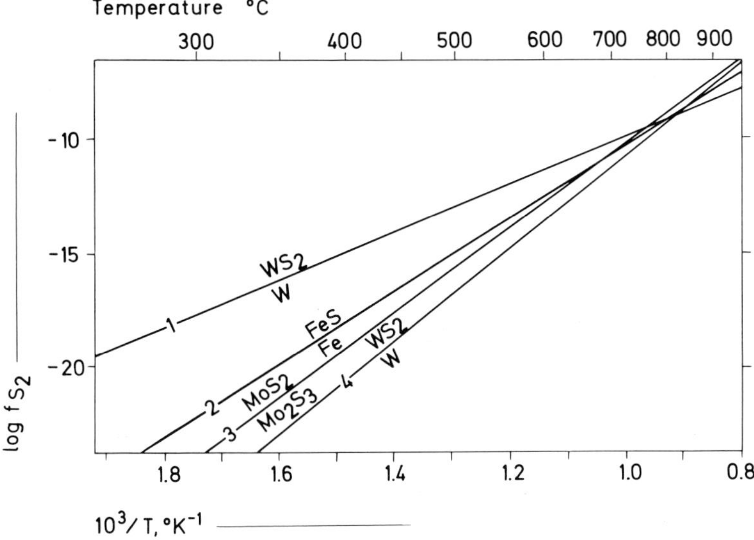

Fig. 16. S_2 activity-temperature diagram of selected sulfidation reactions

nickel-iron phases and troilite (FeS) in meteorites were compared to experimental data[54], yielding three groups of metals, from which a few examples have been selected, namely those included in the present review and study:

1. higher affinity to sulfur than Fe: Mo, Cr, Zn
2. similar affinity to sulfur as Fe: Cu
3. lower affinity to sulfur than Fe: W, Bi.

Thus, tungsten sulfide is rarely found in nature, owing to its distinct affinity to oxygen. Because of the sulfur/oxygen fugacity ratio at ore formation, scheelite ($CaWO_4$) and/or wolframite ($FeWO_4$) occur in nature with pyrrhotite ($Fe_{1-x}S$) and pyrite (FeS_2), rather than tungstenite (WS_2). The phase relations of the Fe–W–S system at 700 °C are shown in Fig. 17. At this temperature pyrrhotite ($Fe_{1-x}S$) can coexist with all stable phases of the system (α-Fe s.s., Fe_2W, Fe_3W_2, W, WS_2, and FeS_2), except liquid sulfur. According to Ref.[6] only Fe_2W and Fe_3W_2 have been confirmed as stable binary compounds in the Fe–W system. Owing to discrepancies in the reported activity measurements (Fig. 16), the univariant W/WS_2 curve No. 4 (taken from Holland[55]) occurs at lower sulfur activity, compared with the Fe/FeS curve No. 2; (data of the Fe/FeS curve No. 2 and the MoS_2/Mo_2S_3 curve No. 3 are given by Barton and Skinner[56]). Both curves, No. 2 and No. 4, intersect at 855 °C. On the other hand, curve No. 2 indicates that FeS is more stable when compared

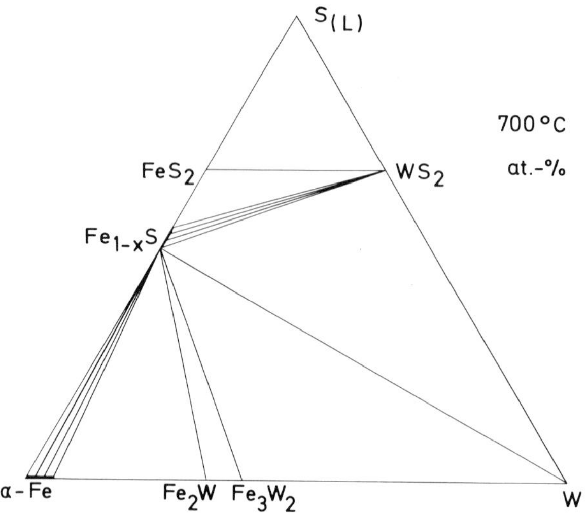

Fig. 17. The 700 °C isotherm of the Fe–W–S system after Stemprok[52)]

with curve No. 1 (data from Kubaschewski and Evans[57)]). These two curves, Nos. 1 and 2, intersect at 799 °C.

Figure 17 demonstrates that FeS and metallic tungsten coexist stably, and experiments show these phase relations exist at least for the whole range of hydrothermal and pneumatolytic-pegmatitic ore deposition. However, with reference to the experimental studies by Vogel and Weizenkorn[53)], no switch of tie lines up to the melting point of the $Fe_{1-x}S$ compound or the appearance of the ternary melt was found.

Tungsten metal was found to coexist with the FeS-rich melt up to 1600 °C.

The following invariant reactions occur within the metal-rich part of the system[53)]:

983 °C — The ternary eutectic occurs;
$FeS + Fe_3W_2 + Fe$ s.s. $\rightleftharpoons$ $liquid_{(E1)} + v$

988 °C — The ternary eutectic melt reaches the binary Fe–S system, the binary eutectic occurs;
$Fe + FeS \rightleftharpoons liquid + v$

993 °C — The ternary peritectic;
$FeS + Fe_3W_2 \rightleftharpoons liquid + W + v$

1087 °C — A second ternary eutectic exists in the univariant region $FeS + WS_2 + W + v$;
$FeS + W + WS_2 \rightleftharpoons liquid_{(E2)} + v$.

The Cu–Mo–S System

The reactions within the ternary Cu–Mo–S system show similarities to the Fe–Mo–S system described above. The main difference is within the Cu–Mo system, where no alloys are formed. The solid solubility of Mo in Cu at 900 °C is „vanishingly small"[6], while ~1.5 wt.% (2.2 at.%) Cu is soluble in Mo at 950 °C[7]. Liquid immiscibility occurs in the complete system at high temperatures.

The binary phase relations of the Cu–S system are complex and have repeatedly been investigated by various researchers. Covellite, stoichiometric CuS, melts incongruently at 507 °C[58] to liquid sulfur and the sulfur-richest endmember of the digenite ($Cu_{1.8}S$)-chalcocite (Cu_2S) solid solution series. This cubic $Cu_{2-x}S$ series is complete above 435 °C; with increasing temperatures it narrows and finally melts at 1129 °C with ~20.3 wt.% S as a singular point[9]. The very complex low-temperature phase relations of the system and related mineral associations are described by Roseboom[59], Morimoto and Koto[60], and Moh[61].

The Cu–Mo–S system at 700 °C, Fig. 18, is deduced from Grover and Moh[62]. At this temperature all stable phases, with the exception of sulfur, are solids. These

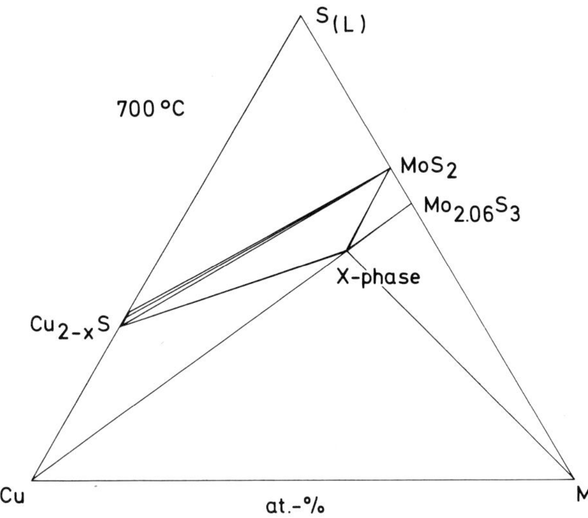

Fig. 18. Phase relations of the Cu–Mo–S system at 700 °C

stable phases are: MoS_2, $Mo_{2.06}S_3$, Mo, Cu, $Cu_{2-x}S$ (s.s.), and a ternary phase, approximately $CuMo_2S_3$. with a limited solid solution range. This ternary phase was first reported as X-phase by Grover[63]; recent single crystal studies have shown rhombohedral symmetry[47].

Divariant relations between $Cu_{2-x}S$ (s.s.) and MoS_2 split the system into sulfur-rich and sulfur-poor portions. The X-phase in the latter region coexists stably with all other phases. At 700 °C (Fig. 18), the X-phase coexists with Cu_2S; however, below 685 ± 5 °C tie lines have switched and metallic copper coexists with MoS_2.

The X-phase can be synthesized above 594 ± 4 °C directly from the elements, or it forms at elevated temperatures in a reaction between copper and molybdenum sesquisulfide:

$$Cu + Mo_{2.06}S_3 \longrightarrow CuMo_{2.06}S_3$$ [62)]

The X-phase is quenchable and remains metastable at room temperature. The lattice parameters are:

$a_{rh} = 6.57$ Å, $\alpha_{rh} = 95°34'$, $V = 278.7$ Å^3 (rhombohedral setting)
or
$a_h = 9.73$ Å, and $c_h = 10.22$ Å, with axial ratio
$a_h/c_h = 0.953$ (hexagonal setting)[47)]

The X-phase is similar to the Y-phase (see Fe–Mo–S system above). With increasing temperatures the X-phase expands its compositional range[5, 62)] and surpasses the Y-phase solid solution width. But in contrast to the Y-phase, the reference composition of the X-phase has at all temperatures a Cu : Mo ratio of 1 : 2.

The ternary eutectic of the Cu–Mo–S system was found to occur at 1063 °C. At this temperature the X-phase shows its maximum solid solution towards the Cu-rich portion of the system which diminishes above. However, with still increasing temperature, the X-phase solid solution width extends towards the Mo–S side, in particular towards MoS_2 or sulfur, respectively. Figure 19 shows a schematic ternary *T*-X block diagram. To simplify the perspective drawing only a few isotherms are indicated. The extension of the whole solid solution width of the X-phase is projected on the base. Thus, the variation and the shifting of the temperature-dependent

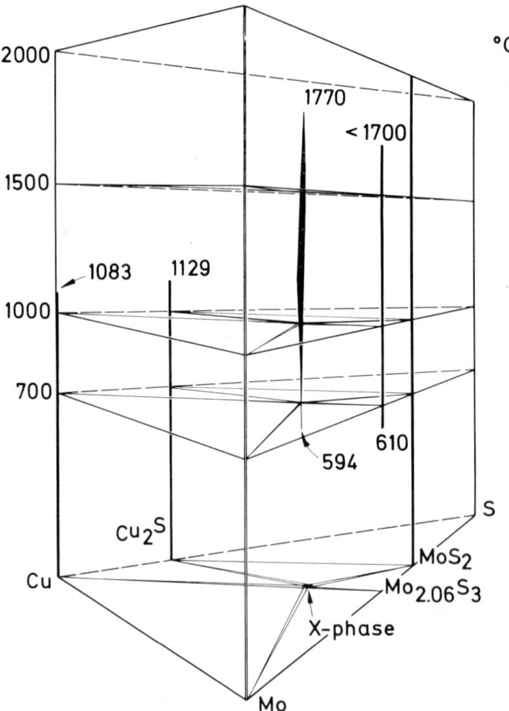

Fig. 19. Schematic T–X_1–X_2 diagram of the Cu–Mo–S system. Temperature stability and solid solution widths of the X-phase (~$CuMo_2S_3$) in the Cu–Mo–S system are displayed. To simplify the perspective drawing a few isotherms are indicated with a projection onto the base, demonstrating the whole compositional X-phase solid solution

solid solution width becomes visible. The figure also indicates the stability ranges of the appropriate binary phases (*e.g.*, MoS_2 is stable throughout the whole temperature range).

The invariant reactions with increasing temperatures in the presence of vapor are as follows (the reactions are only schematic, and equations are not balanced):

500 °C	Since the X-phase is unstable, molybdenite (MoS_2) coexists with all the phases of the system, and divariant relations exist in the diagram with MoS_2 and Cu, MoS_2 and $Cu_{2-x}S$ (s.s.), and with MoS_2 and CuS.
507 °C	Covellite (CuS) melts incongruently in the binary Cu–S system; $CuS \rightleftharpoons S_{(liquid)} + {\sim}Cu_{1.8}S + v$
594 ± 4 °C	The X-phase appears; $Cu + Mo + MoS_2 \rightleftharpoons$ X-phase + v
610 ± 5 °C	$Mo_{2.06}S_3$ becomes stable in the Mo–S binary system between Mo and MoS_2, above which it coexists with the X-phase.
685 ± 5 °C	Copper and molybdenite become unstable; a switch of tie lines: $Cu + MoS_2 \rightleftharpoons Cu_2S$ + X-phase + v
813 °C	The binary monotectic liquid appears in the sulfur-rich portion of the Cu–S system; liquid immiscibility occurs above; ${\sim}Cu_{1.8}S + S_{(liquid)} \rightleftharpoons$ monotectic$_{(liquid)}$ + v; analogous DTA experiments with added MoS_2 contents showed no change of this temperature.

The presence of the X-phase influences the reactions on the metal-rich portion of the system in the following manner:

1063 °C	The ternary eutectic liquid appears in the Cu–Mo–S system; $Cu + Cu_2S$ + X-phase $\rightleftharpoons$ ternary eutectic$_{(liquid)}$ + v
1067 °C	The ternary melt reaches the binary Cu–S system, the binary eutectic occurs; $Cu + Cu_2S \rightleftharpoons$ liquid + v
1102 °C	A ternary monotectic liquid appears in the metal-rich portion of the system; liquid$_{(Cu\text{-}rich)}$ + Cu_2S + X-phase $\rightleftharpoons$ ternary monotectic$_{(liquid)}$ + v
1105 °C	The ternary monotectic melt reaches the binary Cu–S system, the binary monotectic occurs; liquid$_{(Cu\text{-}rich)}$ + $Cu_2S \rightleftharpoons$ binary monotectic$_{(liquid)}$ + v

Above 1105 °C	Liquid immiscibility occurs to a large extent on the copper-rich side of the Cu–Mo–S system.
1080 °C	DTA experiments show that above 1080 °C liquid already appears on the $Cu_{2-x}S$–MoS_2 join
1126 °C	Liquid appears on the join $Cu_{2-x}S$–X-phase.
1129 °C	Is the highest melting point of chalcocite solid solution, $Cu_{2-x}S$, (singular point)[6, 9].

In DTA experimentation the X-phase shows polymorphism (see similarity with the Y-phase, $FeMo_{4.12}S_6$, above)[5, 47]. From 594 ± 4 °C up to ~930 °C, the X-phase appears in its low-temperature γ-modification.

930 ± 10 °C	The γ-modification inverts into the β-polymorph;
1710 ± 10 °C	The β-form inverts into another high-temperature α-modification;
1770 ± 10 °C	Is the melting temperature of the α-form.

Products chilled from ~1800 °C show a slight breakdown, probably due to some vapor loss (Fig. 20): the liquid partly separated into a second Cu_2S-rich melt and metallic molybdenum.

The reported mineral castaingite, $CuMo_2S_{5-x}$[64, 65], probably stable at low temperatures, could not be synthesized in the pure dry system[62].

The Cu–W–S System

The reactions in the Cu–W–S system, though similar, are simpler than those in the other systems above. For geologic reasons Moh[66] has studied the system from near room temperature up to 900 °C. No ternary compounds were found. Thus, the phases which will enter into the ternary reactions are limited to the binaries. Copper and tungsten coexist at all temperatures and no binary alloys occur.

The considerably lower affinity of W (as compared to Cu) for S influences the ternary relations. Figure 21 shows the 700 °C isotherm. The metal-rich portion is dominated by the univariant assembly, Cu + W + Cu_2S + vapor. The $Cu_{2-x}S$ solid solution series produces a bivariant stability region with tungstenite (WS_2). Above 813 °C a wide region of liquid immiscibility occurs in the Cu–S system[9]. WS_2 added to the monotectic composition, *via* DTA and quenching experimentation, did not yield a temperature decrease or show the existence of a ternary monotectic[66]. However, DTA experiments in the *Cu-rich* portion have shown the following thermal effects:

1065 °C	The ternary eutectic appears; $Cu + Cu_2S + W \rightleftharpoons$ liquid + v.

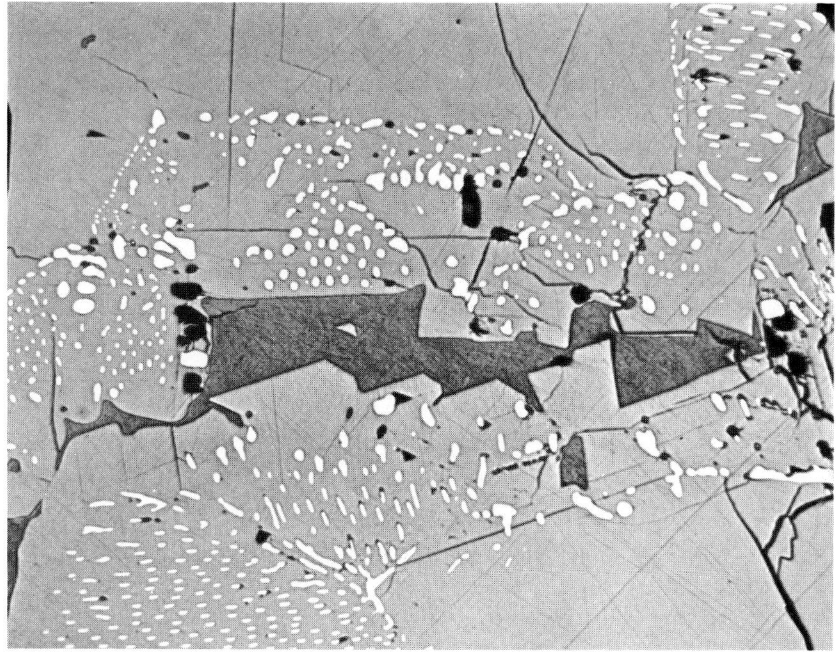

Fig. 20. Breakdown products of melted $CuMo_{2.06}S_3$, rapidly cooled from ~ 1800 °C. A second liquid of approx. Cu_2S composition (*dark grey*) has separated at high temperature from the sulfidic Mo-rich charge (*grey*). Also molybdenum (*white*) has exsolved; oil immersion, x 630

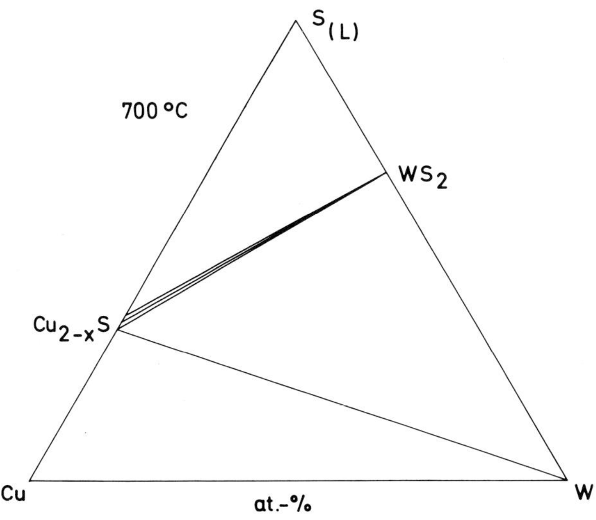

Fig. 21. Phase relations of the Cu–W–S system at 700 °C

1067 °C	The ternary eutectic melt reaches the Cu–S system, the binary eutectic forms; $Cu + Cu_2S \rightleftharpoons$ liquid + v.
1103 °C	A ternary monotectic liquid occurs; $liquid_{(Cu\text{-}rich)} + Cu_2S + W \rightleftharpoons$ ternary $monotectic_{(liquid)}$ + v.
1105 °C	The ternary monotectic melt reaches the Cu–S system, the binary monotectic forms; $liquid_{(Cu\text{-}rich)} + Cu_2S \rightleftharpoons$ binary $monotectic_{(liquid)}$ + v; above which liquid immiscibility occurs.
1107 °C	Liquid appears on the join $Cu_{2-x}S$–WS_2.
1126 °C	Liquid appears on the join $Cu_{2-x}S$–W.
1129 °C	Congruent melting of the pure binary $Cu_{2-x}S$ phase (singular point).

The Bi–Mo–S and Bi–W–S Systems

The Bi–Mo–S system was investigated by Stemprok[31)] with respect to mineral occurrences of the "tin-tungsten-molybdenum ore type"[67)]. In some metallogenic provinces, mineral assemblages such as bismuth + bismuthinite (Bi_2S_3) + molybdenite (MoS_2) are frequently found, because bismuth has a much lower affinity to sulfur as compared with molybdenum.

Bismuth and molybdenum do not alloy together[7)]. One compound, Bi_2S_3, forms in the Bi–S binary system[8)]. No ternary compounds are found in the Bi–Mo–S system[31)]. In the Bi–Mo–S ternary at low temperatures the following univariant assemblages appear: Bi + Mo + MoS_2 + v; Bi + MoS_2 + Bi_2S_3 + v; Bi_2S_3 + MoS_2 + S + v. Bismuth melts at 271.3 °C. With increasing temperature, liquid Bi dissolves increasing amounts of Bi_2S_3[8)]; thus, a distinct divariant region occurs between the metallic liquid and MoS_2. At 610 ± 5 °C $Mo_{2.06}S_3$ is stable, and coexists above with Bi. Figure 22 represents the ternary Bi–Mo–S phase relations at 700 °C; at this temperature the metal-rich liquid extends from pure Bi to nearly 45 at.% S. Above 727 °C a large region of liquid immiscibility forms in the sulfur-rich portion of the Bi–S system, the monotectic composition is reported to be ~67 at.% S. Bi_2S_3 melts congruently at 760 ± 5 °C.

Since the natural Bi-bearing mineral associations were mentioned above[67)], the Bi–W–S ternary phase relations will be briefly discussed.

As the necessary sulfur activity data are available, a ternary diagram can easily be constructed. Bismuth has a lower affinity to sulfur than tungsten, thus bismuthinite would immediately react with tungsten:

$$2\,Bi_2S_3 + 3\,W \longrightarrow 4\,Bi + 3\,WS_2$$

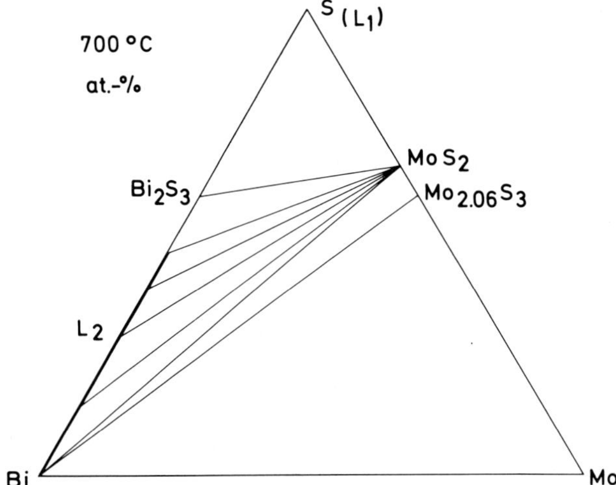

Fig. 22. Phase relations of the Bi–Mo–S system at 700 °C

Consequently, at very low sulfur vapor pressure, Bi + W + WS_2 coexist, whereas with increased sulfur activity the univariant assemblage Bi + WS_2 + Bi_2S_3 becomes stable. At high temperatures WS_2 coexists with a liquid ranging from Bi to beyond Bi_2S_3 composition. No ternary compounds were found to occur in the system.

Additional Ternary Relations

As shown in the above Bi–W–S system, phase relations and reactions can be deduced easily using sulfur activity data from the literature. Other systems are either incomplete or require corrections, for instance when ternary phases occur, which is discussed in the following.

Previously reported ternary (and partly quaternary) Mo-containing sulfidic compounds are becoming more and more interesting, owing to their superconductivity. In 1965, Grover[65)] reported the compound $CuMo_2S_3$ (X-phase); one year later, another compound, ~$FeMo_4S_6$ (Y-phase), was found[44)]. Since then other complex molybdenum sulfides have been reported[48–50)], their characteristic properties measured[68–70)] and methods to synthesize Mo_3S_4 by leaching of preliminary ternary compounds described[21)]. The compositions so far investigated may be represented by a general formual $M_xMo_3S_4$ or $M_xMo_{6-x}S_6$, where M can be any of the following: Ag, Sn, Pb, Mg, Ca, Sr, Ba, Zn, Cd, Cr, Mn, Fe, Co, Ni, Cu, Li, Na, Sc, or Y.

The reported formulas may differ from the experimentally determined ones reported here; for example: $CuMo_2S_3$ (X-phase) and ~$FeMo_4S_6$ (Y-phase), which were obtained as coarse-grained homogeneous materials[62, 46)]. Also the conductivity measurements[68, 69)] may have been influenced by "contamination" of various amounts of the sesquisulfide $Mo_{2.06}S_3$ and excess molybdenum or other phases acting as impurities.

With the large amount of research during the past two decades on the sulfide phase stability relations, their reactions and sulfur activity (compiled by Barton and Skinner[56], Mills[71]), it is now possible with the inclusion of the ternary compounds reported above to construct true phase relations, at least in the metal-rich portion of the Mo-containing systems. A few additional experiments may help to outline the complete phase relations of the systems concerned.

Quaternary Systems

The aim of the present experimental work is the study of reactions and stability relations within multicomponent sulfide systems of geologic interest. Such are: Fe–Mo–W–S, Fe–Cu–Mo–S, Fe–Cu–W–S, and others. Most of the systems have been partially investigated, but slow reaction kinetics, especially in the metal-rich portions, at low temperatures produce disequilibrium assemblages. Thus, at limited reaction times, in some cases the reaction products may appear in such small amounts that they can be overlooked. However, recent experimental investigations have been successfully extended to cover fairly high temperatures.

The Fe–Cu–Mo–S System

The quaternary system, Fe–Cu–Mo–S, is complex. In order to have a complete understanding of this system, all the accessory subsystems must be considered. The Fe–Mo–S and Cu–Mo–S ternary systems have been discussed above. The remaining subsystems are found below.

The Fe–Cu–S system has been completely investigated for its great geologic and technical importance. Of the many contributions only a few are mentioned here. Early workers in synthetic sulfide studies, Merwin and Lombard, in 1937[72], recognized its value. Since then others: Schlegel and Schüller[73] (liquidus and high-temperature relations), Yund and Kullerud[74], and Cabri[75] (entire system at and below 700 °C) have added experimental data. A critical review by Kullerud, Yund and Moh[76] includes new high-temperature results.

Figure 23 shows the system at 700 °C (according to Yund and Kullerud[74]). The abbreviations of phases or mineral names are in accordance with those suggested by Chace[77].

cc = chalcocite, Cu_2S
bn = bornite, Cu_5FeS_4
cp = chalcopyrite, ideal low-temperature composition: $CuFeS_2$
py = pyrite, FeS_2
po = pyrrhotite, $Fe_{1-x}S$

Chalcocite forms a limited binary solid solution series ($Cu_{2-x}S$) extending into the ternary diagram beyond a composition of bornite. The central portion of the system is dominated by the biconvex chalcopyrite stability field. Pyrrhotite displays binary as well as ternary solid solubility, in both cases $>$4 wt.%, whereas pyrite has

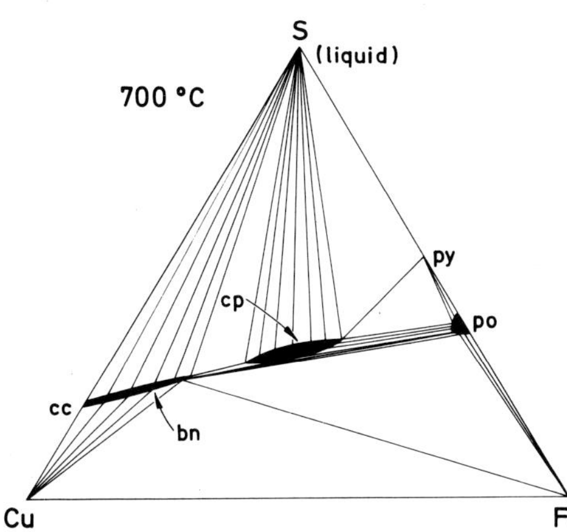

Fig. 23. Phase relations of the Fe–Cu–S system at 700 °C, after Kullerud and Yund[76)]

no measurable solid solution remaining stoichiometric FeS_2. At 700 °C the chalcocite-bornite solid solution series coexists with liquid sulfur, metallic copper; a single member of this solid solution coexists with iron which, at this temperature, dissolves very little Cu. The iron-rich member of the chalcocite-bornite series coexists with the copper-rich portion of the chalcopyrite field and also with a Fe-rich pyrrhotite (~troilite composition). The copper-deficient chalcopyrite field is stable with pyrrhotite and with pyrite. The complete solid solution series forms a divariant region with liquid sulfur.

In the ternary Fe–Cu–Mo system, copper coexists with iron, molybdenum, and with the intermetallic alloy, Fe_3Mo_2.

The quaternary phase relations have been briefly discussed[44, 45)]. The phase relations within the Fe–Cu–Mo–S system are influenced by the complete solid solution series between X-phase (~$CuMo_2S_3$) and Y-phase (~$FeMo_4S_6$), Fig. 24. It is shown as a solid bar. This solid solution series coexists with metallic molybdenum, $Mo_{2.06}S_3$, and MoS_2 (molybdenite). Copper coexists with a Cu-rich member (*i.e.*, pure X-phase) up to a Y-content of approximately 18%. Iron coexists with the solid solution series from ~18 to 75% Y-content, whereas the remainder of the solid solution series forms divariant regions with both Fe and Fe_3Mo_2. Fe_3Mo_2 is also stable with Cu and with an iron-rich member of the bornite solid solution series. A large portion of the X–Y solid solution is in equilibrium with the metal-rich portion of the chalcocite-bornite solid solution series. The more Y-rich members are stable with pyrrhotite or FeS (*i.e.*, troilite).

Molybdenite (MoS_2) forms stable assemblages with the sulfur-rich portion of the chalcocite-bornite solid solution, with chalcopyrite solid solution, with the iron-deficient portion of the pyrrhotite solid solution, and with pyrite. Similar intergrowths between the respective ore minerals, stable at low temperatures, are known from numerous localities.

Series of experiments were performed to outline the high-temperature phase relations of the X–Y phase solid solution series[5)], Fig. 25. Details of the individual ternary reactions have been discussed above. Both endmembers, the X-phase and the Y-phase, stable above 594 ± 4 °C and 535 ± 15 °C, respectively, form a complete solid solution series of pseudo-character, *i.e.*, at low temperatures Cu + MoS_2 +

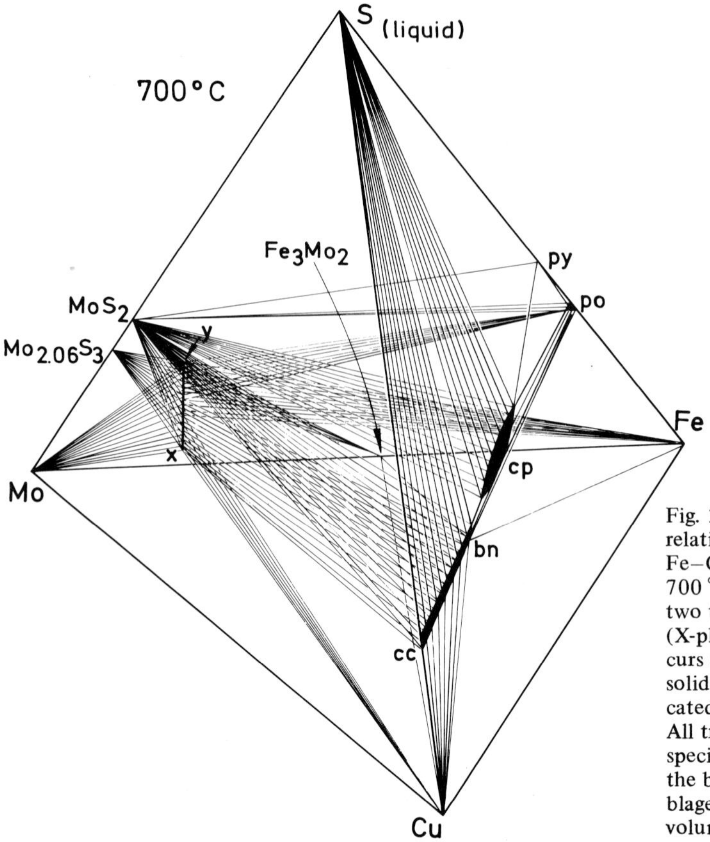

S (liquid)
700 °C
Fe_3Mo_2
py
po
MoS_2
y
$Mo_{2.06}S_3$
Fe
x
Mo
cp
bn
cc
Cu

Fig. 24. Schematic phase relations of the Fe–Cu–Mo–S system at 700 °C. Note: Between the two ternary compounds (X-phase and Y-phase) occurs a complete quaternary solid solution series (indicated simply as a solid bar). All trivariant regions are specified in the text; thus the bivariant phase assemblages and the univariant volumes can be deduced

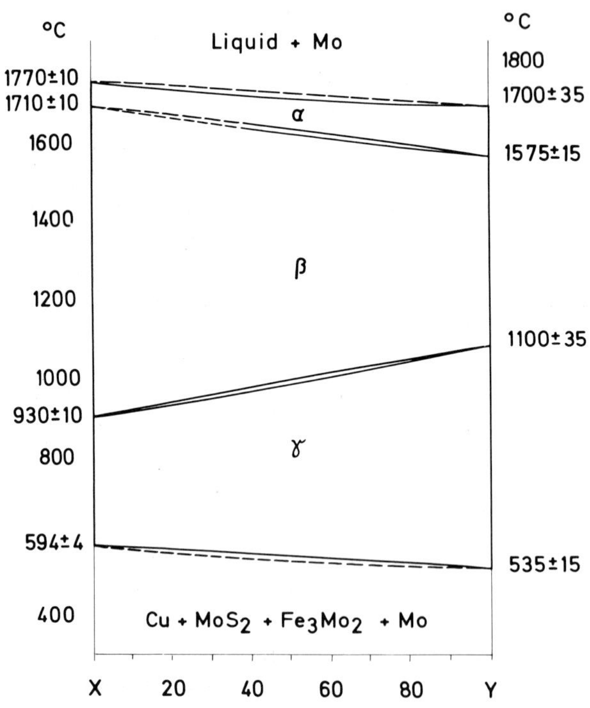

Fig. 25. T–X diagram of the pseudobinary join X-phase --- Y-phase ($\sim CuMo_2S_3$ ----- $FeMo_4S_6$) in weight per cent

Fe_3Mo_2 + Mo coexist. A two-phase region of $\gamma + \beta$ expands from 930 ± 10 °C (X-phase) to 1100 ± 35 °C (Y-phase).

DTA experiments indicated a second transition zone: $\beta + \alpha$ occurs between 1710 ± 10 °C (X-phase) and 1575 ± 15 °C (Y-phase). Finally a continuous liquidus-solidus relation characterizes the high-temperature α-solid solution series with melting temperatures of 1770 ± 10 °C (X-phase) and 1700 ± 35 °C (Y-phase). Melting experiments performed with various X–Y phase mixed crystals resulted in inhomogeneous quenched products. Since a small vapor loss could hardly be prevented, some exsolved molybdenum metal was always observed.

At temperatures $\gtrsim$ 1100 °C the whole X–Y phase solid solution can coexist with a large sulfidic melt, which appears throughout the central portion of the Fe–Cu–S system[76)]. This sulfidic melt reduces the whole quaternary X–Y phase solid solution width, yet the solid solubility increases with increasing temperatures towards MoS_2, in a similar way as for the two endmembers above. The exact shifting of the X–Y phase solid solution range within the quaternary diagram was not determined. However, as found for the endmembers, quenched products of molten intermediate members indicated liquid immiscibility (see Fig. 26). In the X–Y phase matrix a small but visible amount of a (darker) second melt of nearly bornite composition coexisted with metallic Mo.

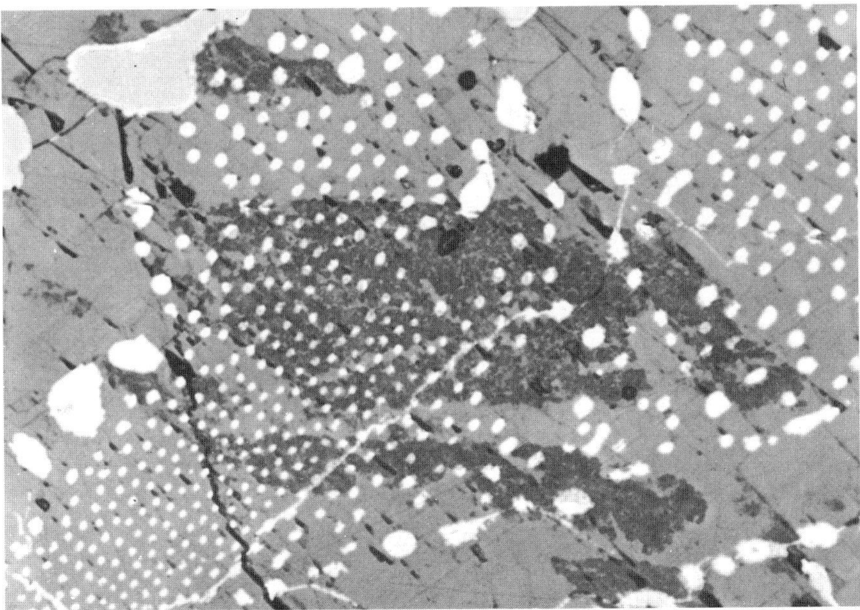

Fig. 26. Previously synthesized homogeneous material of the quaternary X–Y-phase solid solution series (composition ~ $CuFeMo_6S_9$), heated up to ~ 1800 °C, and the melt regulus cooled to room temperature in less than 15 min. The figure shows relics of a darker Cu–Fe-rich sulfidic phase (approx. bornite composition) which separated from the Mo-rich sulfide. Thus, a large region of liquid immiscibility must exist throughout the quaternary system (cf. Figs. 15 and 20). Owing to some vapor loss molybdenum (*white*) has exsolved and remains in small droplets oriented in the sulfidic groundmass. Oil immersion, x 2500

As mentioned earlier, the crystallization from the liquid state along the X-Y-phase solid solution series shows a phenomenon at and near the iron-rich endmember of the series[5]: these cooled melt reguli commonly display turbulence, which could have taken place during crystallization, resulting in outgrowths breaking through the already solidified surface. It is unknown whether this is due to an increased crystallization pressure, or whether it is caused by the establishment of a magnetic field from the heating current (~ 2000 amps). Parallel to the lines of the magnetic force, the "outgrowths" could have been pulled out of the not yet completely crystallized melt. Figure 27 shows this effect on a solidified regulus, which is ground and polished on one side to illustrate the coarse-grained texture. Analogous phenomena were not observed on the more copper-rich parts of the X-Y phase solid solution series.

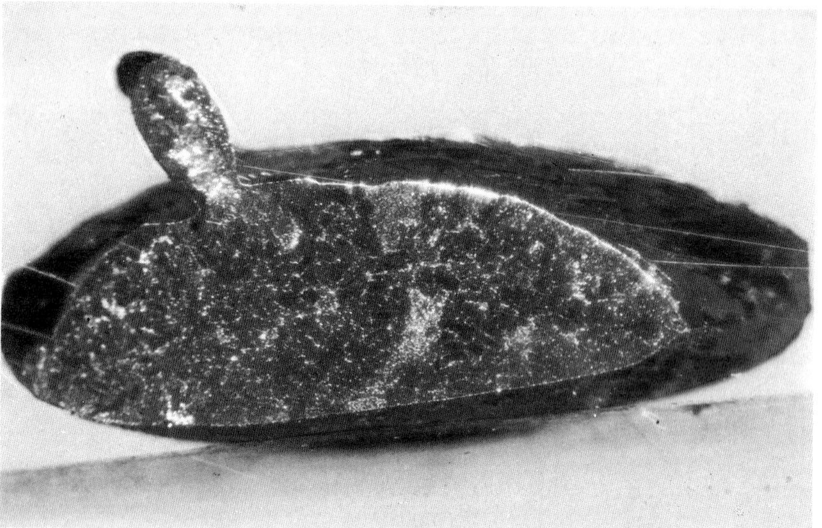

Fig. 27. $FeMo_4S_6$ melt regulus with outgrowths formed during solidification. The regulus, 5 mm diameter, was cut on one side and polished in order to make visible the coarse-grained crystallized sulfidic matrix

Crystallographic studies of the quenched products of the completely miscible compounds $CuMo_2S_3$ and $FeMo_4S_6$ show that the low-temperature γ-polymorphs crystallize in the rhombohedral system with lattice parameters of: $a = 6.57$ Å, $\alpha = 95°34'$, $V = 278.7$ Å^3 for the X-phase endmember, and $a = 6.47$ Å, $\alpha = 94°39'$, $V = 268.7$ Å^3 for the Y-phase endmember[47]. These crystallographic data are in good agreement with other findings: X-phase[68], Y-phase[50].

One of the two high-temperature polymorphs of the Y-phase endmember crystallizes in the triclinic system, with a unit cell closely related to the low-temperature rhombohedral cell. The lattice parameters of the high-temperature polymorph are:
$a = 6.480$ Å, $b = 6.474$ Å, $c = 6.503$ Å,
$\alpha = 91°09'$, $\beta = 97°52'$, $\gamma = 96°12'$, $V = 268.5$ Å^3

Indexed powder data of the $CuMo_2S_3$-$FeMo_4S_6$ mixed-crystal phases are reported by Wang and Moh[47].

The Fe–Cu–W–S System

The relations are comparatively simple within the Fe–Cu–W–S quaternary system[78]. The 700 °C isotherm is shown in Fig. 28, from which the high-temperature equilibria can easily be deduced. Since tungsten does not form ternary or quaternary sulfides, the phase relations of the Fe–Cu–S ternary system dominate and influence the whole tetrahedron. Tungsten is stable only with the very metal-rich portions of the pyrrhotite solid solution (~ FeS composition) and of the chalcocite-bornite mixed-crystal row, as a result of its lower affinity to sulfur. Tungstenite (WS_2) can

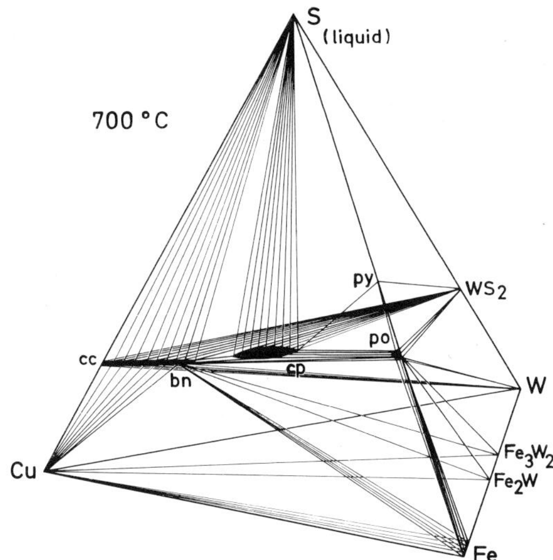

Fig. 28. Schematic phase diagram of the Fe–Cu–W–S system at 700 °C

coexist with the other sulfides: with the chalcocite-bornite s.s., with the whole chalcopyrite stability field, with pyrrhotite and, of course, with pyrite. This is in accordance with natural occurrences[18, 19, 66].

At 700 °C, an iron-rich member of the chalcocite-bornite solid solution coexists with Cu, Fe s. s., and with the alloys Fe_2W and Fe_3W_2. On cooling in the 500 to 475 °C region some quaternary reactions take place, due to the switch of tie lines within the Cu–Fe–S system at 475 ± 5 °C[74]:

bornite + Fe + vapor ⇌ Cu + FeS;

below, the univariant volumes are stable:
FeS + Fe_2W + Fe s.s. + Cu + vapor:
FeS + Fe_2W + Fe_3W_2 + Cu + vapor.

Zn–Mo–W–S Phase Equilibria

In a recent publication zinc was considered for reactions with molybdenum and tungsten sulfides[4]. When compared with the elements discussed above, zinc has the highest affinity to sulfur, and if the pure elements, *e.g.*, Mo or W, Zn and S are allowed to react with each other, ZnS forms immediately, prior to the formation of $Mo_{2.06}S_3$, MoS_2, or WS_2, respectively. Zinc is so aggressive that molybdenum or tungsten sulfides are reduced at once at elevated temperatures,

$$MoS_2 + 2\,Zn \rightarrow Mo + 2\,ZnS$$
$$WS_2 + 2\,Zn \rightarrow W + 2\,ZnS$$

These facts were successfully used to produce Mo–W alloys directly from the corresponding disulfide mixed crystals:

$$(Mo, W)S_2 + 2\,Zn \rightarrow (Mo, W) + 2\,ZnS$$

Reaction kinetics within the pure Mo–W–S system are extremely sluggish[37]. But as the extraction of sulfur by zinc from the disulfide solid solution series is much faster than the disulfides can equilibrate, the step rule is surpassed (*Ostwaldsche Stufenregel*). These phenomena were described in detail by Moh[4].

With the aim of producing homogeneous intermetallic Mo–W solid solution members, the ternary and quaternary reactions referred to above have been used to outline the systems Zn–Mo–S, Zn–W–S, and Zn–Mo–W–S: sphalerite (ZnS) can coexist with all other phases, and no reactions or meltings have been observed between ZnS and Mo, $Mo_{2.06}S_3$, MoS_2, W, WS_2 up to nearly 1400 °C. The substitution of Mo or W for Zn in zinc sulfide is negligibly low and cannot be measured with our experimental methods. On heating, ZnS passes through its transition temperature: cubic ZnS (sphalerite) inverts into a hexagonal α-modification (wurtzite) at 1013 °C[79]. Its melting point is reported to be 1830 ± 20 °C under a pressure of 150 psi[80].

When comparing all these findings with literature data, the occurrence of a ternary zinc-molybdenum sulfide has been reported by Chevrel *et al.*[50]: $ZnMo_nS_{n+1}$; $n = 3, 4, 5, 6$.

Thus, three experiments were performed using pure elements as starting materials and weighings of 1) $ZnMo_3S_4$, 2) $ZnMo_5S_6$, 3) $ZnMo_6S_7$. The charges were heated to 1000 and 1100 °C, reground and then reheated altogether for 9 days, but no homogeneous compounds formed. Polished sections of each experiment definitely show a mixture of three phases: in all cases an excess of molybdenum metal was observed, and a gray-brownish sulfide with optical similarities to molybdenum sesquisulfide ($Mo_{2.06}S_3$), and an excess of ZnS in samples 1) and 2), whereas 3) shows the formation of a strongly pleochroic and anisotropic sulfide, distinctly different from, *e. g.*, $Mo_{2.06}S_3$.

Two additional experiments at 1100 °C were placed on the compositional line between $Mo_{2.06}S_3$ and Zn (similar to the Y-phase composition described above): 4) $ZnMo_{4.12}S_6$, 5) $Zn_{0.8}Mo_{4.12}S_6$. The investigation of the quenched products at room temperature yielded two-phase assemblages in both cases, but no metal excess was observed: gray-brownish sulfide with some ZnS intergrowths. X-ray diffraction patterns of the products indicated similarities to the X–Y phase series for the gray-brownish sulfidic phase.

However, the composition and temperature stability range of this ternary zinc-molybdenum sulfide, which is stable at elevated temperature within the Mo–$Mo_{2.06}S_3$–ZnS triangle of the ternary Zn–Mo–S system, is still obscure. This phase can also coexist with tungsten. Completely unknown are the stability relations of the strongly anisotropic compound which partly formed in experiment No. 3.

The investigations are being continued and a revised phase diagram will be published after completion.

Discussion

The outline of binary, ternary, and quaternary sulfide systems with the VI-B group elements (*e. g.*, Cr, Mo, W) is not claimed to be complete. However, it does represent the current status of research.

The research was investigated by mineralogists to explain further the terrestrial and extraterrestrial occurrence and the mutual reactions of the involved minerals. Since then metallurgists have examined these sulfide phase relations with new processes to extract the valuable metals. The temperature ranges are or can be assumed to be higher than the "geologically relevant temperature range". Using newly developed techniques, laboratory investigations up to 2000 °C have been performed without the reactions actually finishing. This is demonstrated by virtue of the melting reactions of the phases: molybdenite (MoS_2), tungstenite (WS_2), and systems in which they are contained.

Experimental problems still remain to be solved. The "simple" laboratory techniques for high-temperature research discussed above as yet lack a method to balance the sulfur vapor pressure in a "closed system".

The binary sulfide systems discussed in this study are mostly "complete". Apart from the upper P–T stability of the γ-phase, the Cr–S system is completely described.

The Mo–S system is complete up to and including MoS_2. The remaining portion can be performed with time and patience (see discussion above). In addition, most of the metal-rich reactions have been found. Refinement of the high-temperature T–X studies have meanwhile been published[81)].

While the W–S system is shown schematically, recent research work predicts the existence of a high-temperature two-liquid region in the metal-rich portion. This and other results are included in the Annual Report[81)].

These subsystems contain phases of technological interest: MoS_2 is used as a high-temperature lubricant, other suitable refractory sulfides with layered structures might be substituted. However, it has been experimentally proven that some M S_2 compounds will not react with some nonferrous alloys up to 850 °C, even under high pressure.

The completed binary systems facilitate research of the ternary phases and/or ternary solid solutions. Ternary and quaternary molybdenum sulfides have recently been found to have interesting superconductivity properties. The X-phase ($\sim CuMo_2S_3$) and Y-phase ($\sim FeMo_4S_6$) may be only the first of such complex sulfides to be isolated, as greater attention is paid to the chemical preparation of such compounds of Mo-containing sulfide systems. A reported Cd-containing complex molybdenum sulfide shows a transition temperature at 2.5 K, while tin and lead compounds showed $\sim$ 11 K and 13 K, respectively[68]. Transition temperatures may be influenced by the substitution and/or addition of the second element by one of the V-B group (*e. g.*, Nb, Ta) or III-A group (*e. g.*, Al, Ga) forming composition conjugates. For example, the transition temperature increased from $\sim$ 11 K for $SnMo_5S_6$ to 14.4 K for the quaternary $SnAl_{0.5}Mo_5S_6$ compound[69].

It has been well established that "quaternary compounds" are in reality members of solid solution series. In the Fe–Cu–Mo–S system the composition $CuFe_{0.5}Mo_4S_6$ is such a selected "compound", situated between $CuMo_2S_3$ and $FeMo_4S_6$, first reported by Grover and Moh[44].

Mo-containing complex sulfide minerals are rare: femolite ($FeMo_5S_{11}$) and castaingite ($CuMo_2S_5$) have not been synthesized in dry-system experiments[46, 62]. The mineral hemusite (Cu_6SnMoS_8) has been synthesized hydrothermally between 300 and 400 °C[82].

Molybdenum and tungsten have practically identical atomic radii (1.39 Å, based on a coordination of 12). In many compounds and alloys they may substitute for each other. However, tungsten does not form a sesquisulfide. The same atomic radius and similar geochemical properties are possessed by rhenium. As dissolved Re-concentrations (*e. g.*, up to 4000 ppm) were found in molybdenite and tungstenite of the Kipushi ores[66], the introduction of rhenium into further laboratory experimentation is also of geologic relevance. W (and to a certain extent Mo) is found to substitute for Fe in sphalerite-structure minerals found in Cu-rich sulfide ore deposits[83]. The formation of metal-rich ternary sulfides is restricted to molybdenum, since analogous W-compounds have not yet been synthesized.

Certain experimental difficulties (*e. g.*, high temperatures, low reaction rates) have prevented the production of homogeneous compounds[37, 46, 62]. These problems still remain, but new insight has been gained via the simple experimental techniques referred to in the text.

Acknowledgement. Some of the above-mentioned studies have been supported by the Deutsche Forschungsgemeinschaft. This help is greatly appreciated.

References

1) Morey, G. W.: The solubility of solids in gases. Econ. Geol. *52*, 225–251 (1957)

2) Kullerud, G.: Experimental techniques in dry sulfide research. In: Research techniques for high pressure and high temperature (ed. G. C. Ulmer). New York: Springer 1971, pp. 289–315

3) Kullerud, G.: Differential thermal analysis. Carnegie Institution of Washington, Year Book *58*, 161–163 (1959)

4) Moh, G. H.: Experimental and descriptive ore mineralogy, Report 1976. N. Jb. Miner. Abh. *128*, 2, 115–188 (1976)

5) Moh, G. H., Grover, B., Hüller, R.: Experimentelle Untersuchungen an hochtemperaturbeständigen Kupfer-Molybdän- und Eisen-Molybdän-Sulfiden und ihren Mischkristallen. Chemiker-Ztg. *99*, 285–291 (1975)

6) Hansen, M., Anderko, K.: Constitution of binary alloys. 2nd edit. New York: McGraw-Hill 1958, pp. 1305

7) Elliott, R. P.: Constitution of binary alloys. 1 st suppl. New York: McGraw-Hill 1965, pp. 875

8) Shunk, F. A.: Constitution of binary alloys. 2nd suppl. New York: McGraw-Hill 1969, pp. 720

9) Kullerud, G.: Review and evaluation of recent research on geologically significant sulfide-type systems. Fortschr. Miner. *41*, 2, 221–270 (1964)

10) Jellinek, F.: Sulphides. In: Inorganic sulphur chemistry (ed. G. Nickless). Amsterdam-London-New York: Elsevier 1968, pp. 669–747

11) El Goresy, A., Kullerud, G.: Meteorite minerals; the Cr–S and Fe–Cr–S system. Carnegie Institution of Washington, Year Book *67*, 182–187 (1969)

12) El Goresy, A., Kullerud, G.: Phase relations in the system Cr–Fe–S (ed. P. M. Millmann). Meteorite Research No. *53*, 638–656 (1968)

13) Jellinek, F.: The structures of the chromium sulfides. Acta Cryst. *10*, 620–628 (1957)

14) Kullerud, G.: Sulfide phase relations. 50th anniv. Sympos.: Mineralogy and Geochemistry of non-marine evaporites. Miner. Soc. Amer., Spec. public. No. *3*, 199–210 (1970)

15) Bell, R. E., Herfert, R. E.: Preparation and characterization of a new crystalline form of molybdenum disulfide. J. Amer. Chem. Soc. *79*, 3351–55 (1957)

16) Jellinek, F., Brauer, C., Mueller, H.: Molybdenum and niobium sulfides. Nature *185*, 376–377 (1960)

17) Takeuchi, Y., Nowacki, W.: Detailed crystal structure of rhombohedral MoS_2 and systematic deduction of possible polytypes of molybdenite. Schweiz. Mineral. Petrog. Mitt. *44*, 105–120 (1964)

18) Graeser, S.: Über Funde der neuen rhomboedrischen MoS_2-Modifikation (Molybdänit-3R) und von Tungstenit in den Alpen. Schweiz. Mineral. Petrog. Mitt. *44*, 121–128 (1964)

19) Ramdohr, P.: Die Erzmineralien und ihre Verwachsungen. 4. Aufl. Berlin: Akademie-Verlag 1975, pp. 1277

20) Gmelins Handbuch der anorganischen Chemie: Molybdän-Wolfram-Uran; 8th edit. System No. *53*, Berlin: Verlag Chemie 1935, pp. 182–188

21) Chevrel, R., Sergent, M., Prigent, J.: Un nouveau sulfure de Molybdene: Mo_3S_4 preparation, propriétés et structure cristalline. Mat. Res. Bull. *9*, 11, 1487–98 (1974)

22) Morimoto, N., Kullerud, G.: The Mo–S system. Carnegie Institution of Washington, Year Book *61*, 143–144 (1962)

23) Stubbles, J. R., Richardson, F. D.: Equilibrium in the system molybdenum + sulfur + hydrogen. Trans. Faraday Soc. *56*, 1460–66 (1960)

24) Jellinek, F.: Structure of molybdenum sesquisulfide. Nature *192*, 1065–66 (1961)

25) Semiletov, S. A.: The crystalline structure of rhombohedral MoS_2. Soviet Phys. Cryst. *6*, 428–431 (1962)

26) Wilderwanck, J. C., Jellinek, F.: Preparation and crystallinity of molybdenum and tungsten sulfides. Z. Anorg. Chem. *328*, 309–318 (1964)

27) Moh, G. H., Udubasa, G., Hüller, R.: Hochtemperatur-Phasengleichgewichte im System Molybdän-Schwefel. Metall *28*, 8, 804 (1974)

28) Dreizler, W., Moh, G. H.: High temperature DTA experimentation, unpublished research, Stuttgart (1976)

29) Cannon, P.: Melting point and sublimation of molybdenum disulfide. Nature *183*, 1612–13 (1959)

30) Takeno, S., Moh, G. H.: High temperature experimentation with selected sulfides, unpublished research, Hiroshima/Japan (1976)

31) Štemprok, M.: The Bi–Mo–S system. Carnegie Institution of Washington, Year Book *65*, 336–337 (1967)

32) Clark, A. H.: Evidence for the geological occurrence of molybdenum trisulfide. Nature Phys. Science *234*, 177–178 (1971)

33) Rode, E. Ya., Lebedev, B. A.: Physico-chemical investigations on molybdenum trisulfide and on its thermal break-down products. (Orig. in Russian) Zh. Neorgan. Khim. *6*, 1189–1197 (1961)
34) Wells, R. C., Butler, B. C.: Tungestenite, a new mineral. J. Wash. Acad. Sci. *7*, 596–599 (1917)
35) Samoilov, S. M., Rubinstein, A. M.: Study of physical and chemical properties of WS_2 catalysator. (Orig. in Russian) Izv. Akad. Nauk. SSSR, Otd. Khim. Nauk. *11*, 11905–11912 (1959)
36) Ehrlich, P.: Untersuchungen an Wolframsulfiden. Z. Anorg. Chem. *257*, 247–253 (1948)
37) Moh, G. H., Udubasa, G.: Molybdänit-Tungstenit-Mischkristalle und Phasenrelationen im System Mo–W–S. Chem. Erde *35*, 327–335 (1976)
38) Agte, C., Vacek, J.: Wolfram a Molybden. Prague: Staatsverl. f. techn. Literatur, 1954; German translation by H. Herklotz: Wolfram und Molybdän. Berlin: Akademie-Verl. 1959
39) Höll, R., Weber-Diefenbach, K.: Tungstenit-Molybdänit-Mischphasen in der Scheelitlagerstätte Felbertal (Hohe Tauern, Österreich). N. Jb. Miner. Mh. H. *1*, 27–34 (1973)
40) Kullerud, G., Yoder, H. S.: Pyrite stability relations in the Fe–S system. Econ. Geol. *54*, 4, 533–572 (1959)
41) Power, L. F., Fine, H. A.: The iron-sulphur system. Minerals Sci. Engng. (Johannesburg, S. A.) Vol. *8*, 2, 106–128 (1976)
42) Ramdohr, P.: Der Mundrabilla-Meteorit. Fortschr. Miner. *53*, 2, 165–186 (1976).
43) Bell, P. M., El Goresy, A., England, J. L., Kullerud, G.: Pressure-temperature diagram for Cr_2FeS_4. Carnegie Institution of Washington, Year Book *68*, 277–278 (1970)
44) Grover, B., Moh, G. H.: Experimentelle Untersuchungen des quaternären Systems Kupfer-Eisen-Molybdän-Schwefel. Vortrag DMG-Tagung, München 1966, Referate (1966)
45) Moh, G. H.: Experimental sulfide petrology and its application. Izvj. Jugoslav. centr. krist. (Zagreb) *10*, 5–14 (1975)
46) Grover, B., Kullerud, G., Moh, G. H.: Phasengleichgewichtsbeziehungen im ternären System Fe–Mo–S in Relation zu natürlichen Mineralien und Erzlagerstätten. N. Jb. Miner. Abh. *124*, 3, 246–272 (1975)
47) Wang, N., Moh, G. H.: Crystallographic and microscopic studies on the refractory sulfides: $\sim CuMo_2S_3$ and $FeMo_4S_6$. N. Jb. Miner. Mh. H. *1*, 36–43 (1976)
48) Chevrel, R., Sergent, M.: Chimie minérale. Preparation de thiomolybdites d'éléments de transition. C. R. Acad. Sc. Paris *267*, 1135–36 (1968)
49) Guillevic, J., Chevrel, R., Sergent, M.: Étude radiocristallographique de $FeMo_2S_4$ et de $CoMo_2S_4$. Bull. Soc. Fr. Mineral. Cristallogr. *93*, 495–97 (1970)
50) Chevrel, R., Sergent, M., Prigent, J.: Sur de nouvelles phases sulfurées ternaires du molybdène. Jour. Solid State Chem. *3*, 515–519 (1971)
51) Skvorcova, K. V., Sidorenko, G. A., Dara, A. D., Silantjeva, I. I., Medoeva, M. M.: Femolite a new sulphide of molybdenum. Zapiski vses. mineral obshch. *93*, 436–443 (1964)
52) Štemprok, M.: The iron-tungsten-sulphur system and its geological application. Miner. Deposita *6*, 302–312 (1971)
53) Vogel, R., Weizenkorn, H.: Über das Dreistoffsystem Eisen-Schwefel-Wolfram. Archiv Eisenhüttenwesen *32*, 6, 413–420 (1961)
54) Vogel, R.: Ergebnisse ternärer Zustandsdiagramme des Eisens, angewandt auf Fragen der Kosmochemie. Chem. Erde *21*, 24–47 (1962)
55) Holland, H. D.: Some applications of thermochemical data to problems of ore deposits. II. Mineral assemblages and the composition of ore-forming fluids. Econ. Geol. *60*, 1101–1166 (1965)
56) Barton, P. B., Skinner, B. J.: Sulfide mineral stabilities. In: Geochemistry of hydrothermal ore deposits (ed. H. L. Barnes). New York: Holt, Rinehardt & Winston 1967, pp. 236–333
57) Kubaschewski, O., Evans, E.: Metallurgical thermochemistry. (Russ. translation); London: Butterworth-Springer (1951)
58) Kullerud, G.: Covellite stability relations in the Cu–S system. Freiburger Forschungsh. *C 186*, 145–160 (1965)
59) Roseboom, E. H.: An investigation of the system Cu–S and some natural copper sulfides between 25° and 700 °C. Econ. Geol. *61*, 641–672 (1966)

60) Morimoto, N., Koto, K.: Phase relations of the Cu–S system at low temperatures: stability of anilite. Am. Mineralogist *55*, 106–117 (1970)
61) Moh, G. H.: Blue remaining covellite and its relations to phases in the sulfur-rich portion of the copper-sulfur system at low temperatures. Mineral. Soc. Japan (Proc. IMA-IAGOD Meetings '70, IMA-Vol.) spec. pap. 1, 226–232 (1971)
62) Grover, B., Moh, G. H.: Phasengleichgewichtsbeziehungen im System Cu–Mo–S in Relation zu natürlichen Mineralien. N. Jb. Miner. Mh. 529–544 (1969)
63) Grover, B.: New phase equilibria in the system copper-molybdenum-sulphur. N. Jb. Miner. Mh. H. *7*, 219–221 (1965)
64) Schüller, A., Ottemann, J.: Castaingit, ein neues mit Hilfe der Elektronen-Mikrosonde bestimmtes Mineral aus dem Mansfelder „Rücken". N. Jb. Miner. Abh. *100*, 317–321 (1963)
65) Clark, A. H., Sillitoe, R. H.: Supergene $CuMo_2S_5$ („Castaingite"), Potrerillos, Atacama Province, Chile. N. Jb. Miner. Mh. 499–503 (1969)
66) Moh, G. H.: Das Cu–W–System und seine Mineralien sowie ein neues Tungstenitvorkommen in Kipushi/Katanga. Mineral. Deposita *8*, 291–300 (1973)
67) Cissarz, A.: Übergangslagerstätten innerhalb der intrusivmagmatischen Abfolge. N. Jb. Miner. Geol. Paleontol. *56*, 99–274 (1928)
68) Matthias, B. T., Marezio, M., Corenzwit, E., Cooper, A. S., Barz, H. E.: High-temperature superconductors, the first ternary system. Science *175*, No. 4029, 1465–66 (1972)
69) Fischer, Ø., Odermatt, R., Bongi, G., Jones, H., Chevrel, R., Sergent, M.: On the superconductivity in the ternary molybdenum sulfides. Physics Letters *45A*, 2, 87–88 (1973)
70) Marezio, M., Dernier, P. D., Remeika, J. P., Corenzwit, E., Matthias, B. T.: Superconductivity of ternary sulfides and the structure of $PbMo_6S_8$. Mat. Res. Bull. *8*, 657–668 (1973)
71) Mills, K. C.: Thermodynamic data for inorganic sulfides, selenides and tellurides. London: Butterworths 1974
72) Merwin, H. E., Lombard, R. H.: The system Cu–Fe–S. Econ. Geol. *32*, 203–284 (1937)
73) Schlegel, H., Schüller, A.: Die Schmelz- und Kristallisationsgleichgewichte im System Kupfer-Eisen-Schwefel und ihre Bedeutung für die Kupfergewinnung. Freiberger Forschungsh., B, Hüttenwesen, Metallurgie *2*, 31 (1952)
74) Yund, R. A., Kullerud, G.: Thermal stability of assemblages in the Cu–Fe–S system. Jour. Petrology *7*, 3, 454–488 (1966)
75) Cabri, L. J.: New data on phase relations in the Cu–Fe–S system. Econ. Geol. *68*, 4, 443–454 (1973)
76) Kullerud, G., Yund, R. A., Moh, G. H.: Phase relations in the Cu–Fe–S, Cu–Ni–S, and Fe–Ni–S systems. Econ. Geol. Monograph *4*, 323–343 (1969)
77) Chace, F. M.: Abbreviations in field and mine geological mapping. Econ. Geol. *51*, 712–723 (1956)
78) Moh, G. H.: Ore mineral systems; Report 1975. N. Jb. Miner. Abh. *126*, 2, 126–145 (1976)
79) Moh, G. H.: The sphalerite-wurtzite inversion, pp. 38–40. In: Tin-containing mineral systems, Part. II. Phase relations and mineral assemblages in the Cu–Fe–Zn–Sn–S system. Chem. Erde *34*, 1, 1–61 (1975)
80) Addamiano, A., Dell, P. A.: The melting point of zinc sulfide. J. Phys. Chem. *61*, 1020–1021 (1957)
81) Moh, G. H.: Ore minerals; Report 1977. N. Jb. Miner. Abh.: N. Jb. Miner. Abh. *131*, 1, 1–55 (1977)
82) Nekrasov, I. Ya.: Phase relations of hemusite in the system Cu–Mo–Sn–S between 300 and 400 °C. In: Phase relations in tin-containing systems. (Orig. in Russian). Moscow: Akad. Nauk Verl. 1976, pp. 179–186
83) Moh, G. H.: Scheme of mineral systems: microscopical observations on typical sulfide ores with an interpretation based on laboratory experiments. Fortschr. Miner. *55*, 1, 79–104 (1977)

Received May 12, 1977

Author Index Volumes 26–76

The volume numbers are printed in italics

NMR

Basic Principles and Progress
Grundlagen und Fortschritte

Editors: P. Diehl, E. Fluck, R. Kosfeld

Volume 9
Lyotropic Liquid Crystals
1975. 18 figures, 3 tables. IV, 85 pages
ISBN 3-540-07303-5
Contents: Nuclear Magnetic Resonance Studies in Lyotropic Liquid Crystals: Introduction. Studies of Lyotropic Liquid Crystals. Studies of Molecular and Ionic Species Dissolved in the Nematic Phase of Lyotropic Liquid Crystals.

Volume 10
Van der Waals Forces and Shielding Effects
1975. 13 figures, 46 tables. II, 118 pages
ISBN 3-540-07340-X
Contents: Historical Development (up to 1961). Continuum Models. Pair Interaction Models. Other Experimental Proton Data on σ_W. The Physical Nature of the Field F^2 and of the Associated Excitation Energy. The Site Factor. The Repulsion Effect. The Effect of Higher Order Dispersion Terms. The Parameters B. σ_W in Dense Media. The Temperature Dependence of σ_W. Factor Analysis. ^{19}F σ_W Studies. σ_W of Nuclei other than ^1H and ^{19}F. Alternative Referencing Systems. On the Required Molecular Parameters and Physical Constants.

Volume 11
M. Mehring
High Resolution NMR Spectroscopy in Solids
1976. 104 figures. XI, 246 pages
ISBN 3-540-07704-9
Contents: Introduction. Nuclear Spin Interactions in Solids. Multiple-Pulse NMR Experiments. Double Resonance Experiments. Magnetic Shielding Tensor. Spin-Lattice Relaxation in Line Narrowing Experiments. Appendix.

Volume 12
B. Lindman, S. Forsén
Chlorine, Bromine and Iodine NMR Physico-Chemical and Biological Applications
1976. 74 figures, 29 tables. XIV, 368 pages
ISBN 3-540-07725-1
Contents: Introductory Aspects. Relaxation in Molecules or Ions with Covalently Bonded Halogens. Shielding Effects in Covalent Halogen Compounds. Scalar Spin Couplings. Relaxation of Chloride, Bromide and Iodide Ions. Shielding of Halide Ions. Quadrupole Splittings in Liquid Crystals. Halide Ions in Biological Systems. Studies of the Perchlorate Ion.

Volume 13
Introductory Essays
Editor: M. M. Pintar
1976. 48 figures. XI, 154 pages
ISBN 3-540-07754-5
Contents: A Guide to Relaxation Theory. Thermodynamics of Spin Systems in Solids. Coherent Averaging and Double Resonance in Solids. Macroscopic Dipole Coherence Phenomena. Nuclear Spins and Non Resonant Electromagnetic Phenomena. Nuclear Spin Relaxation in Molecular Hydrogen. Longitudinal Nuclear Spin Relaxation Time Measurements in Molecular Gases. Spin-Lattice Relaxation in Nematic Liquid Crystals. Via the Modulation of the Intramolecular Dipolar Interactions by Order Fluctuations. Studies of Molecular Tunnelling. Effect of Molecular Tunnelling on NMR Absorption and Relaxation in Solids. How to Build a Fourier Transform NMR Spectrometer for Biochemical Applications.

Volume 14
H. Nöth, B. Wrackmeyer
Nuclear Magnetic Resonance Spectroscopy of Boron Compounds
1978. 1 figure, 96 tables. XII, 461 pages
ISBN 3-540-08456-8
Contents: Introduction and Scope. Nuclear Magnetic Properties of Boron. ^{11}B NMR or Two-Coordinate Boron. ^{11}B Chemical Shifts of Three Coordinate Boron. ^{11}B NMR of Transition Metal Boron Compounds. ^{11}B NMR of Diborane and Derivatives. ^{11}B NMR of Tetracoordinate Boron. Spin-Spin Coupling Constants $^nJ(^{11}BX)$. Tables of ^{11}B NMR Data. References. Author Index.

Volume 15
H. W. Spiess, A. Steigel
Dynamic NMR Spectroscopy
1978. 76 figures. VI, 213 pages
ISBN 3-540-08784-2
Contents: Rotation of Molecules and Nuclear Spin Relaxation. Mechanistic Studies of Rearrangements and Exchange Reactions by Dynamic NMR Spectroscopy.

Springer-Verlag
Berlin Heidelberg New York